Lecture Notes in Mathematics

Edited by A. Dold, B. Eckmann and F. Takens

1458

Renate Schaaf

Global Solution Branches of Two Point Boundary Value Problems

Springer-Verlag
Berlin Heidelberg New York London
Paris Tokyo Hong Kong Barcelona

Author

Renate Schaaf
Utah State University, Department of Mathematics and Statistics
Logan, UT 84322-3900, USA

Mathematics Subject Classification (1980): 34B15, 35B32, 34C25

ISBN 3-540-53514-4 Springer-Verlag Berlin Heidelberg New York
ISBN 0-387-53514-4 Springer-Verlag New York Berlin Heidelberg

© Springer-Verlag Berlin Heidelberg 1990
Printed in Germany

Printing and binding: Druckhaus Beltz, Hemsbach/Bergstr.
2146/3140-543210 – Printed on acid-free paper

Introduction

For the parameter dependent problem

(I-1-1)
$$u''(x) + \lambda^2 f(u(x)) = 0 \quad , \quad \lambda > 0$$
$$u(0) = u(1) = 0$$

the following is known from application of general local ([13]) or global ([24]) bifurcation theorems in the case $f(0) = 0$, $f'(0) > 0$:

The trivial solution $u \equiv 0$ exists for each λ. From this trivial solution branch there is bifurcation of nontrivial solutions in each point $(0, \lambda)$ for which the linearized problem (I-1-1) has a nontrivial kernel, i.e., for $\lambda = i\pi/\sqrt{f'(0)}$, $i = 1, 2, \dots$. The bifurcating branches, i.e., connected components of nontrivial solutions with bifurcation points in the (λ, u)-space, are all unbounded and each branch consists of solutions (λ, u) where u has a number of simple zeroes in $]0, 1[$ which is characteristic for the branch. In the solution branches bifurcating from 0 all u are bounded by the first positive and first negative zero of f.

In applications more information about the shape of solution branches is needed. It is easy to see that all branches are in fact curves which are at least as smooth as f is (see below). It is then of interest whether these curves have turns with respect to the λ-direction or not, and if so, how many turning points there are and where they are located. Let, e.g., the branch of positive and negative solutions to (I-1-1) look like this:

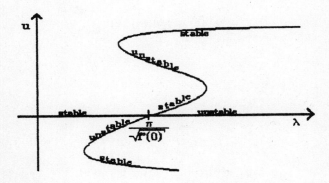

Figure I.1.1

Then (I-1-1) is the stationary equation of

(I-1-2)
$$u_t = u_{xx} + \lambda^2 f(u)$$
$$u(t,0) = u(t,1) = 0$$

and the directions of the solution branches determine the stability of the stationary states as indicated.

In combustion problems there often occurs an equation of the form (I-1-1) describing intermediate steady states of the temperature distribution u, λ then measures the amount of unburnt substance. In this context turning points of a branch correspond to ignition and extinction points of the process, and it is of importance whether or not those exist (see e.g.[15]).

In some cases one can skilfully choose sub- and supersolutions of (I-1-1) and then get the result that there exist at least two stable solutions for a certain λ-interval. Then using degree theory there has to be at least one more unstable solution for any λ in this interval. This way it is shown that turning points of the branch have to exist, but one only gets estimates from below for both the number of solutions and the number of turns of the branch. It is not possible to give upper estimates of these numbers without using strong analytical tools.

We use for this purpose the so called time map T (see [31]) of the nonlinearity f:

First of all by a scaling of x, $x = \lambda t$ we can write (I-1-1) in the form

(I-1-3)
$$u''(t) + f(u(t)) = 0$$
$$u(0) = u(\lambda) = 0$$

thus having the parameter in the boundary condition.

If $u(t) = U(t,p)$ is the solution of the initial value problem

(I-1-4)
$$u'' + f(u) = 0$$
$$u(0) = 0, \; u'(0) = p \neq 0$$

then we define $T(p) = T_1(p)$ to be the first positive t for which $U(t,p)$ is zero again, if this one exists for the given p:

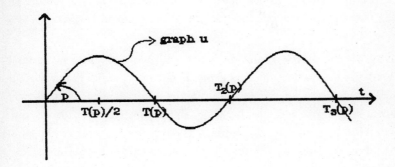

Figure I.1.2

$T_2(p)$ is the second zero of u, $T_3(p)$ the third (existence provided) etc..

Then the solution branches of (I-1-3) are given by the graphs of the T_i:

(λ, u) solves (I-1-3) if and only if $u = U(\cdot, p)$, $T_i(p)$ is defined for some $i = 1, 2, ...$
and

(I-1-5)
$$\lambda = T_i(p).$$

Stable regions of positive or negative solution branches are $u'(0) = p$-regions with $pT'(p) > 0$, unstable ones correspond to $pT'(p) < 0$, turning points of branches correspond to $T'(p_0) = 0$, $T''(p_0) \neq 0$.

Time maps and period maps (see below) have been studied in several papers, see [3], [5], [8], [9], [22], [25], [26], [31], [33], [35], [36], a list of references which is by no means complete.

By the implicit function theorem T_i is always at least as regular as f is since

$$U(T_i(p), p) = 0 \qquad \left| \frac{\partial}{\partial t} U(T_i(p), p) \right| = |p| > 0.$$

The last identity follows since (I-1-4) has a first integral

(I-1-6)
$$\frac{1}{2}(u')^2 + F(u) \equiv const = \frac{1}{2}p^2$$

for $u = U(\cdot, p)$ and F being the integral of f with $F(0) = 0$.

Note that with T and T_i we get all solutions of (I-1-3), not only the branches bifurcating from 0. Also it is not necessary that we have a trivial solution set via $f(0) = 0$. (There are general results for global branches in these cases too, we did not mention them.)

From (I-1-6) one can derive a formula for T which is a singular integral (see (1-1-3)). For studying derivatives of T this one is not very useful. Therefore in section 1.1 we derive a time map formula for branches bifurcating from zero which does not contain a singularity and can be easily differentiated. The idea behind this is to use a transformation in u which maps the orbits of (I-1-4) in the phase plane into circles. For solutions in bifurcating branches this can be done since f does not change sign more than once along the range of such solutions.

Chapter 1 consists entirely of applications of this time map formula. Section 1.2 recalls results about bifurcation points and bifurcation directions. In 1.3 we use the time map formula to reprove results by Chafee, Infante ([5]) and Opial ([22]) giving conditions for branches which have only a single turn at the bifurcation point itself.

Paragraph 1.4 goes a step further and gives conditions on f under which T_i'' does not change sign. This results in branches which still have at most one turning point, but this one can occur away from the bifurcation point (see figure I.1.2). This allows for a change of stability along branches, but stability can change at most twice.

We have tried to find a condition on f for $T_i'' \neq 0$ which can be verified in examples. For this we introduce the notion of an A-B-function which is a function f with

(I-1-7)
$$f'f''' - \frac{5}{3}(f'')^2 < 0 \quad \text{in regions where } f' \geq 0.$$
$$ff'' - 3(f')^2 < 0 \quad \text{in regions where } f' \leq 0.$$

It is shown that $T_i'' > 0$ for all bifurcating branches of A-B-functions. This class contains all polynomials without complex zeroes as is shown in 1.5. With this we have generalized the result of Smoller and Wasserman in [31] which says that the T_i have at most one critical point which then is a minimum in the case that f is a cubic $f(u) = -u(u - \alpha)(u - \beta)$ with $\alpha < 0 < \beta$. 1.5 contains more methods for getting hold of A-B-functions, using, e.g., the Schwarzian of f. See also the examples discussed near the end of this introduction.

Section 1.6 deals with the asymptotic behaviour of T as p approaches a boundary point of its definition set. Some of the results can be used to prove existence of dead core solutions to certain reaction diffusion problems (see examples 1.6.2 and 3.2.2).

Chapter 2 contains results on the Neumann problem corresponding to (I-1-3)

(I-1-8)
$$u'' + f(u) = 0$$
$$u'(0) = u'(\lambda) = 0\,,$$

and related results on period maps of Hamiltonian systems

(I-1-9)
$$u' = f_2(v) \qquad v' = -f_1(u)\,.$$

If r is some zero of f then $u \equiv r$ is always a trivial solution of (I-1-8). The linearization of (I-1-8) about $u \equiv r$ has a nonzero solution if and only if $f'(r) > 0$, $\lambda = i\pi/\sqrt{f'(r)}$, $i = 1, 2, \dots$. These are the bifurcation points of (I-1-8). Solutions u in the bifurcating branches are bounded by the first zeroes of f below and above r. For these branches one can show that they have a single turn at the bifurcation point in the case that f is an A-B-function. The proof uses a Neumann time map defined analogously to the Dirichlet time map mentioned before.

With about the same method we get monotonicity results for the period map of (I-1-9), which is the map assigning to E the least period $\Pi(E)$ of a periodic solution of (I-1-9) with energy $E \equiv F_1(u) + F_2(v)$. Such results can be used to prove bifurcation of subharmonic solutions if (I-1-9) is perturbed by a small nonautonomous t-periodic term (see [7]). Another application of such results is the existence and uniqueness of period-4-solutions to the time delay equation $u'(t) + f(u(t-1)) = 0$ with antisymmetric f (see [18]).

In chapter 3 we return to the Dirichlet problem and give results for problem (I-1-3) where we assume less for f and also consider non bifurcating branches:

In 3.1 we no longer assume $f(0) = 0$, $f'(0) > 0$ but just consider some f defined on $]0, a[$ which is positive there. This way we can handle the branch of (I-1-3) which consists of solutions (λ, u) with u positive and u'' not changing sign. For these we get a time map formula similar to the one in 1.1.

The regularity of f required for the existence of T turns out to be a fairly weak one. Therefore problems with singularities at $u = 0$ can be dealt with. Multiple zeroes of f at 0 can also occur. In this case there is "bifurcation" from $(\lambda = \infty, u \equiv 0)$. If f is an A-B-function then the branch is shown to be again U-shaped.

The case $f(0) > 0$ shows more difficulties than $f(0) = 0$: For $f(u) = -(u - \alpha)(u - \beta)(u - \gamma)$ with $\alpha < \beta < 0 < \gamma$ it is shown in [31] that the number of critical points of T is at most two, and exactly two if β is close to 0. So far we are not able to generalize this to A-B-functions f, though it seems like this result is true. The only thing we could do is to give a criterion on f such that T has at most one critical point. In the case $f(0) > 0$ it is not possible to show this via a sign of T'' or a related expression. Instead we use variation diminishing properties of integral operators proved in [19], results are to be found in 3.3.

In 3.4 we discuss positive solution branches in which u'' changes sign, i.e. u has at least one zero of f in its range. Under certain assumptions on f we can again show that branches are U-shaped.

The last chapter discusses some general features of Dirichlet time maps. In 4.1 we give proofs for the relationship between the sign of $pT'(p)$ and stability. Also there is a strong connection between $T(p)$ and the energy level of the corresponding solution of (I-1-3) if we define the energy level via the Liapounov functional

$$E(u) = \int_0^\lambda \frac{1}{2}(u'(t))^2 - F(u(t))\, dt .$$

It turns out that E is decreasing along parts of solution branches with $pT'(p) > 0$ and increasing if $pT'(p) < 0$. As another result if there are multiple solutions for a fixed λ an unstable solution always has higher energy than the two "nearest" stable solutions.

In section 4.2 we again only consider nonlinearities f defined on some interval $]0, a[$ which are positive there. We pose the inverse problem if it is possible to decide for a given curve $p \mapsto T(p)$ whether or not it is the time map of some f and if it is possible to calculate f from T, i.e., to get the problem back from its solution branch. The set of all time maps of such f is characterized by an integral condition and the operator $f \mapsto T$ is shown to be invertible.

Applications include proofs of generic properties of such time maps analogous to [3], [33], as well as estimates for λ-regions of existence and nonexistence for (I-1-3), and can be found in 4.3 among some games about possible and impossible time maps. In the appendix, finally, we briefly discuss the method to obtain time map plots via the computer.

In order to give a flavour of the possible applications results in this monograph can have let us discuss two specific examples. Please note that the argumentation used to interpret results is not rigorous up to the detail.

It is a good illustration to view (I-1-1) as the stationary equation for a population of size u diffusing on the interval $[0,1]$ which has a hostile environment thus forcing $u(t,0) = u(t,1) = 0$. $1/\lambda^2$ can then be regarded to be the diffusion coefficient of the population, wheras $f^+(u) = f(u)/u$ models the reproduction rate consisting of the birth rate minus the death rate.

Let us first consider the example

$$f^+(u) = e^{-(u-\alpha)^2} \quad , \quad \alpha \geq 0.$$

This means that the reproduction rate is maximal for a certain optimal population size α and declines to 0 as u becomes large but the death rate never exceeds the birth rate. This means for the model without diffusion

$$u_t = u f^+(u)$$

that any population with a positive intitial size grows unlimited. If we now add diffusion

(I-1-10)
$$u_t = \frac{1}{\lambda^2} u_{xx} + u f^+(u)$$
$$u(t,0) = u(t,1) = 0$$

then the behaviour depends on the size of λ and is in principal governed by the bifurcation diagram for the stationary equation (I-1-1).

All stationary nontrivial branches are given by $\lambda = T_i(p)$. In this specific context only the positive solution branch $\lambda = T(p)$, $p > 0$, matters since the system is not going to get near any sign changing steady state if it starts out with $u(0,x) \geq 0$ (maximum principle). Section 1.2 gives the bifurcation points $(\lambda_i, u \equiv 0)$ as

$$\lambda_i = T_i(0) = \frac{i\pi}{\sqrt{f'(0)}} = i\pi\sqrt{e^{\alpha^2}}$$

and the bifurcation directions as

$$T_i'(0) = \begin{cases} 0, & i \text{ even} \\ -\dfrac{2}{3}\dfrac{f''}{(f')^2}(0) = -\dfrac{2}{3}\alpha e^{\alpha^2}, & i \text{ odd} \end{cases}.$$

So the positive branch starts out unstable if $\alpha > 0$.

For the "other end" of the branches we use section 1.6: First of all the definition set of the T_i is (see section 1.1) $D(T_1) =]b^-, b^+[$ with $b^- = -\sqrt{2F(-\infty)}$, $b^+ = \sqrt{2F(\infty)}$ and $D(T_i) =]b^-, -b^-[$ for $i \geq 2$.

From proposition 1.6.1 (iv) it follows that $\lim_{p \to b+} T(p) = \frac{\pi}{\sqrt{c^+}}$ with $c^+ = \lim_{u \to \infty} \frac{f^2(u)}{2F(u)} = \lim_{u \to \infty} f^+(u) = 0$ and $\lim_{p \to b-} T(p) = \frac{\pi}{\sqrt{c^-}}$ with $c^- = \lim_{u \to -\infty} f^+(u) = 0$. So

$$\lim_{p \to b+, b-} T(p) = +\infty .$$

Formulas (1-6-17) and (1-6-18) then imply for all the other branches that $\lim_{p \to b-, -b-} T_i(p) = +\infty$, $i \geq 2$.

Let us put together what we have found out by this for the positive branch $\lambda = T(p)$, $p > 0$: It starts out unstable but has to recover stability later since $T(p) \to +\infty$ as $p \to b^+$. If we let $\lambda^* = \min_{p \geq 0} T(p) > 0$ then no nontrivial steady state exists for $\lambda < \lambda^*$ and the zero solution is a global attractor for the system ([14], [17]). The population dies out from too fast diffusion and the hostile environment. For $\lambda > \lambda_1 = \pi e^{\alpha^2/2}$ the zero solution looses its stability, but the population could never grow beyond limits: From the Liapounov functional and the fact that $f^+(+\infty) = 0$ it follows that any solution of (I-1-10) is a priori bounded. So diffusion and the zero boundary conditions prevent unlimited growth.

Results in [17] can be used to obtain further information on the dynamical behaviour of our system for $\lambda > \lambda^*$ if we know that all steady states are hyperbolic (this follows from $T_i'(p) \neq 0$ if $\lambda = T_i(p)$) and if we know more about the maximal invariant set A which consists of the unstable manifolds of all steady states present for our λ. For this example this is possible by applying sections 1.3, 1.4, 1.5 and the results about connecting orbits obtained by Brunovský and Fiedler in [4].

Let us first consider the case $\alpha = 0$. Then $u \frac{d}{du} \frac{f(u)}{u} = u f^{+\prime}(u) < 0$ and we can use theorem 1.3.2 to obtain that $p T_i'(p) > 0$, which completely proves the following bifurcation diagram:

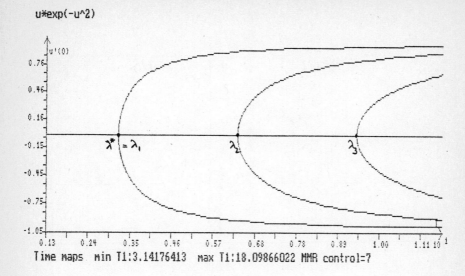

u*exp(-u^2)

Time maps   min T1:3.14176413   max T1:18.09866022 MMR control=7

Figure I.1.3

In this case λ coincides with λ^* and it follows ([14], [17]) that our population dies out for $\lambda < \lambda_1$ and for $\lambda > \lambda_1$ is attracted by the single stable positive steady state represented by $\lambda = T(p)$.

If we do not assume that the maximal reproduction rate occurs at $u = 0$, i.e., if we take $\alpha > 0$, then we know already that $T'(p) < 0$ near $p = 0$. So there is at least one turning point of the positive branch in the nontrivial range. We can show that there is at most one by showing that $T'' > 0$. This again follows from the results in 1.4 if f is an A-B-function (see (I-1-7)). Now it is clear that (I-1-7) follows from $(\ln|f|)'' < 0$ and $(\ln|f'|)'' < 0$ on regions where these are defined and from $ff''(u) < 0$ whenever $f'(u) = 0$. The last and the first condition can be easily checked. For the middle one we notice that

$$f'(u) = e^{-(u-\alpha)^2}(A - u)(u - B)$$

with $A = \frac{1}{2}(\alpha + \sqrt{\alpha^2 + 2})$, $B = \frac{1}{2}(\alpha - \sqrt{\alpha^2 + 2})$, thus

$$(\ln|f'|)''(u) = -2 - \frac{1}{(A - u)^2} - \frac{1}{(u - B)^2} < 0.$$

So we have proved the following bifurcation picture:

u*exp(-(u-0.5)^2)

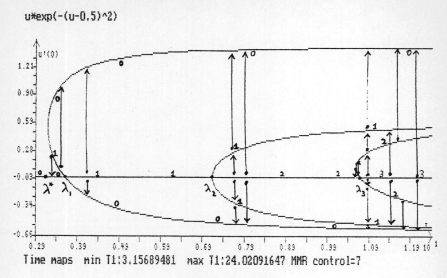

Time maps   min T1:3.15689481   max T1:24.02091647 MMR control=7

Figure I.1.4

This diagram determines the complete qualitative dynamical behaviour of our time dependent system, also if we additionally consider initial conditions which are allowed to be negative in subsets of $]0,1[$: The numbers along the branches indicate the dimension of the unstable manifold of each of the stationary solutions in that part of the branch. In [4] it is proved that for this type of a global bifurcation diagram there is an orbit of (I-1-10) from any given unstable steady state u to each steady state v whoose unstable dimension is lower. These connecting orbits are indicated as arrows. The dimension of the set of orbits connecting u to v is just the difference between their unstable dimensions. Stationary points together with their connecting orbits form the maximal compact invariant set A for the flow to which everything is eventually attracted ([17]). Since we can deduce the flow on A from the bifurcation diagram, the complete qualitative behaviour of (I-1-10) is known. Every initial distribution which does not lie in either one of the stable manifolds of the unstable solutions will be attracted to one of the stable steady states. This set of initial conditions is open and dense in $H^1([0,1])$. If we only consider initial conditions $u(0,x) \geq 0$, $\not\equiv 0$, as is appropriate in our model context then the following can be said: For $\lambda < \lambda^*$ we always have $\lim_{t\to\infty} u(t,x) = 0$. If $\lambda^* < \lambda < \lambda_1$ then there are two positive steady states $u_1 < u_2$ with u_1 being unstable, u_2 stable and with $u \equiv 0$ being an additional stable steady state. Then the unstable manifold of u_1 (a set of codimension 1) divides the set of initial conditions into two components which are then either attracted by u_2 or 0. If $u(0, x) > u_1(x)$ then $\lim_{t \to \infty} u(t, x) = u_2(x)$ whereas for $u(0, x) < u_1(x)$ we get $\lim_{t \to \infty} u(t, x) = 0$, the initial population size was not large enough to allow survival. For $\lambda > \lambda_1$ the zero solution has become unstable and there is a single positive steady state to which any positive initial condition is attracted.

Next it might be interesting to give an estimate of the value λ^* below which everything dies out. We can do this in theory using proposition 4.3.2 in chapter 4:

$$\lambda^* \geq \min_{u \geq 0} \pi \frac{\sqrt{2F(u)}}{f(u)}, \qquad \lambda^* \leq \min_{u \geq 0} \pi \frac{u}{\sqrt{2F(u)}}.$$

But since $F(u)$ cannot be calculated if $\alpha > 0$ we can only give a more crude concrete estimate from below using f': From (4-3-4) we get

$$\min_{u \geq 0} \pi \frac{\sqrt{2F(u)}}{f(u)} \geq \frac{\pi}{\sqrt{\max_{u \geq 0} f'(u)}}.$$

Now $f'(u) = e^{-(u-\alpha)^2}(1 + 2\alpha u - 2u^2) \leq \max_{u \geq 0}(1 + 2\alpha u - 2u^2) = 1 + \alpha^2/2$. So

$$\lambda^* \geq \frac{\pi\sqrt{2}}{\sqrt{2 + \alpha^2}}.$$

In the previous example no unlimited growth in the presence of diffusion was possible because of $f^+(\infty) = 0$. This is different if we choose

$$f^+(u) = \frac{u - 1}{u + 1}.$$

The choice of the production rate carries another difference: It is negative for u near 0, which is an appropriate modelling for a population with sexual reproduction. This way we no longer have $f'(0) > 0$, thus no bifurcation from the trivial solution branch occurs and the the results of section 3.4 for nonbifurcating branches apply. From there we get that the time map for positive solutions $T(p)$ is defined for $p \in]0, \infty[$. Representation (3-4-13) for $T(p)$ together with proposition 3.1.4 shows that $\lim_{p \to \infty} T(p) = \pi\sqrt{c}$ with $c = \lim_{u \to \infty} \frac{2F(u)}{f^2(u)} = \frac{1}{f^+(+\infty)} = 1$ and that $\lim_{p \to 0} T(p) = +\infty$.

In the case that f starts out negative near 0 and only has a single sign change on $]0, \infty[$ condition (3-4-23) for theorem 3.4.4 is always satisfied. This tells us that

(I-1-11) $$\frac{d^2}{d\hat{p}^2} T(p) > 0 \quad \text{with } p^2 = \hat{p}^2 + 2F(1)$$

provided f is an A-B-function where it is positive, that is on $]1, \infty[$. Since f'
is positive there we only have to check the first part of (I-1-7) which follows easily
from $f''' < 0$.

So there is a change of variables for T which turns it into a convex function.
Because of the asymptotic behaviour we then conclude that T' is always negative
and the $\lambda^* = \pi$ is the minimal value of T:

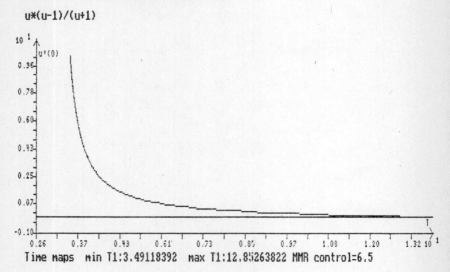

Figure I.1.5

Hence, as in the model without diffusion the $0-$ solution is always stable, but its
region of attraction changes with λ. If the unstable steady state u_1 given by
$\lambda = T(p)$ is present (for $\lambda > \lambda^*$) then any initial condition below u_1 is attracted
to 0 whereas we expect anything above it to grow unlimited.

In this example it is interesting to see the influx of a diffusion rate which depends
on the population density. If we modify our model to

(I-1-12)
$$u_t = \frac{1}{\lambda^2}(d(u)u_x)_x + f(u)$$
$$u(t,0) = u(t,1) = 0$$

with $d(u) > 0$ then the bifurcation diagram of steady states can change with the
form of $d(u)$. With $D(u) := \int_0^u d(s)\,ds$, h being the inverse function of D and
with $u = h(v)$ (I-1-12) becomes equivalent to

(I-1-13)
$$(h(v))_t = \frac{1}{\lambda^2}v_{xx} + f(h(v))$$
$$v(t,0) = v(t,1) = 0.$$

This problem has a stationary equation (I-1-1) with v taking the place of u and $\tilde{f} := f \circ h$ the place of f.

Since in our example $f''' < 0$ we not only have that f is an A-B-function but that the Schwarzian Sf of f (see 1.5) is negative where $f, f' > 0$. Then by lemma 1.5.7 we get that $f \circ h$ is an A-B-function for any h with $Sh \leq 0$. This is in particular true if h is a linear fractional transformation because then $Sh = 0$. So choosing any linear fractional transformation for D will give us the result that T is a convex function modulo a transformation in p. The particular behaviour of T then only depends on the aymptoticts near the boundary of its definition set. Let us choose the two possible examples of D which yield two rather different bifurcation diagrams:

First consider

$$(I\text{-}1\text{-}14) \qquad\qquad D(u) = \frac{\alpha u}{u + \beta} \qquad \alpha, \beta > 0.$$

This implies that the diffusion rate $d(u)$ starts at a positve level at $u = 0$ and decreases to 0 as $u \to \infty$, modelling an effect of increasing stickiness. Then $\tilde{f}$ is defined for $0 \leq v < \alpha$, and the definition set of T is again $]0, \infty[$. Since $\tilde{F}(\alpha) = \tilde{f}(\alpha) = +\infty$ we conclude from 3.1.4 together with (3-4-13) that $T(p) \to 0$ as $p \to \infty$. So this is the bifurcation diagram (here $\alpha = 1$, $\beta = 2$):

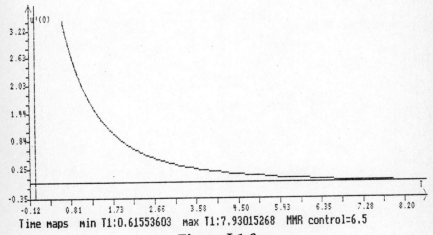

Figure I.1.6

The difference to the situation $d(u) \equiv 1$ is that a "hair trigger" unstable steady state $u_1 = h(v_1)$ is present for all λ. So the strategy of "sticking together" has the effect that the 0 always has a finite region of attraction and that unlimited growth is never excluded. Note that this is possible since the production rate $f^+(u)$ has a positive lower bound as $u \to \infty$ thus we assume infinite resources for the population. There is no way that any choice of $d(u)$ could alter the fact of restricted growth in our first example since there the decrease of f^+ towards 0 is too fast. Actually it is not obvious that the bifurcation diagram here translates into the asymptotic behaviour of (I-1-12) or (I-1-13) in the same way as in the first example. But the reader can easily work out that this is in fact the case using the remarks at the end of section 3.2.

The other possible choice of D is

(I-1-15) $$D(u) = \frac{\alpha u}{\beta - u} \quad , \quad \alpha, \beta > 0.$$

This way (I-1-12) only makes sense for $0 \le u < \beta$ and $d(u)$ increases from a positive level at $u = 0$ to infinity as $u \to \beta$. This could model a phobic reaction if the population density becomes too large. (Don't take the model interpretation too serious, it is only included here to make things a little more illustrative. But these are the lines along which "real" models could be discussed.) $\tilde{f}(v) = f(h(v))$ then is defined for $0 \le v < \infty$ with $\tilde{f}(v)/v \to 0$ as $v \to \infty$ since $h(v) \to \beta$ as $v \to \infty$ and f^+ is bounded. So with the same sort of arguments as before $T(p) \to \infty$ as $p \to \infty$ and $T(p) \to \infty$ as $p \to 0$. This way the following picture is proved (again here $\alpha = 1$ and $\beta = 2$):

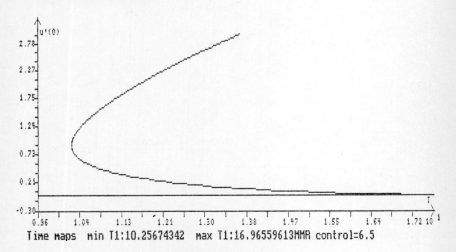

Figure I.1.7

Thus the choice of $d(u)$ has taken us back to the qualitative situation of the first example with the only difference that the region of attraction of the zero solution is always an open bounded set since the hair trigger unstable positive steady state is always present.

These examples have hopefully shown how results proved in this monograph can be applied to give some insight into the dynamical behaviour of model problems as (I-1-10) or (I-1-2). Another direction of applications is the following:

Consider an equation similar to (I-1-10):

(I-1-16) $u_t = \triangle_x u + \lambda^2 f(u) \quad t > 0, \, x \in \,]0, 1[\times \mathbb{R}.$

Then the number of traveling wave solutions of (I-1-16) is closely related to the number of solutions to (I-1-1), which then becomes the equation for the possible starting and end configurations of the traveling wave on the finite cross section of the x-domain of (I-1-16). These results can be found in [16].

As a whole the tools required to apply the results obtained in this work need some more mathematical sophistication than the ones needed to actually prove them. This is why I wrote this paper with the intention that it should be understandable (at least in large parts) by anybody with a good knowledge of calculus and a basic background in the theory of ordinary differential equations. To see the problems in a larger framework some notion of facts from bifurcation theory, partial differential equations and functional analysis come in useful. I hope that more expert readers will find it easy to skip passages with well known facts and find enough interesting material in the rest.

I had the chance to discuss problems about time maps with many people having a related field of work and I am grateful to all for the resulting stimulus for my work. Special thanks go to H.Engler for discussions which showed once again that it can be fun doing mathematics and which opened up new ways of looking at problems, though the fine theorem we had in mind turned out to be wrong (see the beginning of 3.3). Also I want to thank P.Brunovský explicitly who asked me two questions whose answers now make up the better part of chapter 4, and U.Kirchgraber from whom I got the idea to consider more general integrable systems instead of second order equations. Thanks also to J.Smoller and B.Fiedler for encouraging discussions and to the latter, P.Brunovský and S.Heintze for providing the necessary theoretical links for applications. Last but not least let me thank my teacher W.Jäger for sharing his broad knowledge and his views of which problems are interesting problems and which are not.

CHAPTER I

DIRICHLET BRANCHES BIFURCATING FROM ZERO

We want to find out properties of Dirichlet-branches bifurcating from 0. So we consider nonlinearities f with $f(0) = 0$ and $f'(0) > 0$ up to their first zeroes $>$ or < 0. In this chapter we thus impose the following conditions on f:

(1-0-1)
$$f(u) = uf^+(u) \quad \text{with} \quad a^- < 0 < a^+ \quad \text{and}$$
$$f^+ : \,]a^-, a^+[\, \to \mathbb{R}^+ \quad \text{locally Lipschitz-continuous}$$

With this the initial value problem

(1-0-2)
$$u''(t) + f(u(t)) = 0$$
$$u(0) = 0, \, u'(0) = p$$

will always have a unique solution $u(t) =: U(t, p) \not\equiv 0$ for any $p \neq 0$, and $p \mapsto U(t, p)$ will be locally Lipschitz continuous.

We can thus define the Dirichlet time map of (1-0-2) by

(1-0-3)
$$D(T) := \{p \neq 0 \mid U(t, p) = 0 \text{ for some } t > 0\}$$
$$T(p) := \min\{t > 0 \mid U(t, p) = 0\} \quad \text{for } p \in D(T)$$

$T(p)$ will give the first zero of the solution $U(\cdot, p)$. To get solutions of the corresponding boundary value problem which change sign we recursively define for $i = 1, 2, 3, \ldots$

(1-0-4)
$$T_1 := T$$
$$D(T_i) := \{p \in D(T_{i-1}) \mid U(T_{i-1}(p) + t, p) = 0 \text{ for some } t > 0\}$$
$$T_i(p) := T_{i-1}(p) + \min\{t > 0 \mid U(T_{i-1}(p) + t, p) = 0\} \quad \text{for } p \in D(T_i)$$

So $T_i(p)$ is just the i-th zero of $U(\cdot, p)$, if it exists. The functions T_i are locally Lipschitz continuous. Most readers will notice immediately that T_i for $i > 1$ can be calculated from T, but we will not mind this momentarily, since in the following section we will get a uniform representation for all the T_i.

The connection between the time maps and global solution branches of

(1-0-5)
$$u'' + f(u) = 0$$
$$u(0) = u(\lambda) = 0$$

then is the following:

(λ, u) is a solution of (1-0-5) with $u \not\equiv 0$

iff

$u(t) = U(t, p)$ for some $p \in D(T_i)$, $i \in \{1, 2, ...\}$,

and

$\lambda = T_i(p)$.

The solutions of (1-0-5) with u having exactly $i - 1$ zeroes in $]0, \lambda[$ are given by the curve $p \mapsto (T_i(p), U(\cdot, p))$. The graph of T_i is the i-th solution branch of (1-0-5) projected to the $(\lambda, u'(0))$ - plane. $T_i(p)$ for $p > (<) 0$ gives the branch part in which all u are positive (negative) near 0, especially the branch of positive solutions is given by $T(p)$, $p > 0$, and the negative branch by $T(p)$, $p < 0$.

1.1 The time map formula

Let F be the first integral of f, i.e.,

(1-1-1) $$F(u) := \int_0^u f(s)\, ds.$$

Then (1-0-2) can be integrated to give

(1-1-2) $$\frac{1}{2}(u')^2 + F(u) \equiv \frac{1}{2}p^2.$$

With this it is possible to get a first formula for the time map. Since any solution of (1-0-2) is symmetric in a max or min, we have that $T(p)$ is just twice the time the solution needs to get to the first zero of u'. If we call this time t_0 then on $]0, t_0[$ we can make the change of variables $u = u(t)$ with $du = u'(t)\, dt$, i.e., using (1-1-2), $du = \sqrt{p^2 - 2F(u)}\, dt$ for $p > 0$ and $du = -\sqrt{p^2 - 2F(u)}\, dt$ for $p < 0$.

With this we get for $p > 0$ that

$$t_0 = \int_0^{t_0} 1\, dt = \int_0^{F_+^{-1}(\frac{1}{2}p^2)} \frac{du}{\sqrt{p^2 - 2F(u)}},$$

where F_+ is F restricted to the set $]0, a^+[$ where f is positive. For $p < 0$ a corresponding formula holds. So

(1-1-3) $$T(p) = \kappa 2 \int_0^{F_\kappa^{-1}(\frac{1}{2}p^2)} \frac{du}{\sqrt{p^2 - 2F(u)}} \quad \text{with} \quad \begin{cases} \kappa = +\text{ for } p > 0 \\ \kappa = -\text{ for } p < 0 \end{cases}$$

This formula is not very useful since it is not at all obvious how to calculate $T'(p)$ from $T(p)$, though we know it has to exist if f is sufficiently regular.

The way to overcome these problems is to change over to angular coordinates in the phase plane rather than using $u = u(t)$. So one could use $u(t) = r(\theta)\sin\theta$, $u'(t) = r(\theta)\cos\theta$, which is done in [22].

Here we will make use of the Hamiltonian structure of (1-0-2) and introduce angular coordinates which transform (1-0-2) into a linear equation. This way we will get a representation of the time maps in terms of linear integral operators. For a similar transformation see [35], [36].

First, for any f as in (1-0-1) we define a map

(1-1-4)
$$x \mapsto \Phi(f)(x) := g(x) \quad \text{via}$$
$$b^- := -\sqrt{2F(a^-)} \quad \text{and} \quad b^+ := \sqrt{2F(a^+)},$$
$$g : \,]b^-, b^+[\,\rightarrow\,]a^-, a^+[\quad \text{with}$$
$$F(g(x)) := \frac{1}{2}x^2 \quad , \quad \text{sign}g(x) = \text{sign}x.$$

Then g is continuously differentiable in $x \neq 0$, and because of the special assumptions we have made on f in (1-0-1) it is also continuously differentiable in 0:

By implicit differentiation we get

(1-1-5)
$$g'(x) = \frac{x}{f(g(x))} = \frac{\sqrt{2F(u)}}{|f(u)|} > 0 \quad \text{where} \quad u := g(x).$$

thus, since $g(0) = 0$

$$(g'(x))^2 = \frac{2F(u)}{f^2(u)} = \frac{2F(u)}{u^2} \frac{1}{(f^+(u))^2} \rightarrow \frac{1}{f^+(0)} \quad \text{as} \quad u \rightarrow 0$$

by de l'Hôpital's theorem.

Hence

(1-1-6)
$$g(0) = 0 \quad \text{and} \quad g'(0) = \frac{1}{\sqrt{f^+(0)}} > 0.$$

g' also is locally Lipschitz since f^+ is.

We will use g for a transformation of the initial value problem equivalent to (1-0-2)

(1-1-7)
$$u' = y$$
$$y' = -f(u)$$
$$u(0) = 0, \, y(0) = p \neq 0.$$

If we now use the transformation $u = g(x)$ we get that $g'(x)f(u) = x$. So (1-1-7) becomes

(1-1-8)
$$g'(x)x' = y$$
$$g'(x)y' = -x$$
$$x(0) = 0, \, y(0) = p.$$

Now by a subsequent transformation of time (1-1-8) becomes a linear equation:
Since x and g' are sufficiently regular the initial value problem

$$(1\text{-}1\text{-}9) \qquad \frac{dt}{d\theta} = g'(x(t))$$
$$t(0) = 0$$

will uniquely define a function $t(\theta)$ which is continuously differentiable. If we now differentiate with respect to θ rather than t then for (1-1-8) we get

$$(1\text{-}1\text{-}10) \qquad \frac{dx}{d\theta} = y$$
$$\frac{dy}{d\theta} = -x$$
$$x(0) = 0, \, y(0) = p.$$

Thus $x(t(\theta)) = p \sin \theta$, $y(t(\theta)) = p \cos \theta$ and, going back through the transformations

$$(1\text{-}1\text{-}11) \qquad t(\theta) = \int_0^\theta g'(p \sin \varphi) \, d\varphi$$
$$u(t(\theta)) = g(p \sin \theta).$$

u will have a first zero if all values of $p \sin \theta$ with θ between 0 and π are in the definition set of g. This is the case for $p \in \,]b^-, b^+[$. For more than one zero of u all values of $p \sin \theta$ with θ between 0 and 2π have to be in the definition set of g, i.e.,both p and $-p$ have to be in $]b^-, b^+[$. The i-th zero of u will then be $t(i\pi)$.

So we are finally able to write down the time map formula:

1.1.1 THEOREM. *Let (1-0-1) hold for f. Then the following formulas hold for the time maps of f as defined by (1-0-3), (1-0-4):*

$$(1\text{-}1\text{-}12) \qquad D(T) = \,]b^-, b^+[$$
$$T(p) = \int_0^\pi g'(p \sin \theta) \, d\theta$$

with $g = \Phi(f)$, b^- and b^+ defined by (1-1-4).
For $i > 1$

$$(1\text{-}1\text{-}13) \qquad D(T_i) = \,] - \bar{b}, \bar{b}[$$
$$T_i(p) = \int_0^{i\pi} g'(p \sin \theta) \, d\theta$$

with $\bar{b} := \min(b^+, -b^-)$.

If f is sufficiently regular then g' will be too, and formulas (1-1-12), (1-1-13) can be easily differentiated. We will just note this in the next proposition:

1.1.2 PROPOSITION. *Let* f *satisfy (1-0-1), and let* f^+ *be* k-*times continuously differentiable.*

Then $g = \Phi(f)$ *is* $k + 1$-*times continuously differentiable on* $]b^-, b^+[$ *with*

$$(1\text{-}1\text{-}14) \qquad g''(x) = \frac{f^2 - 2Ff'}{f^3}(u) \quad \text{for} \quad k \geq 1,$$

$$(1\text{-}1\text{-}15) \qquad g'''(x) = -g'(x)\frac{3f'(f^2 - 2Ff') + 2Fff''}{f^4}(u) \quad \text{for} \quad k \geq 2,$$

where u *stands for* $g(x)$.

T_i *then is* k-*times continuously differentiable for* $i = 1, 2 \ldots$ *with*

$$(1\text{-}1\text{-}16) \qquad T_i^{(n)}(p) = \int_0^{i\pi} \sin^n(\theta)\, g^{(n+1)}(p\sin\theta)\, d\theta.$$

PROOF: By (1-1-5)

$$(g'(x))^2 = \frac{2F(u)}{u^2}\frac{1}{(f^+(u))^2}, \quad u = g(x).$$

Using the Taylor formula for f^+ about 0 one can see that $(2F(u))/(u^2)$ is k-times continuously differentiable with respect to u. The same holds for $1/f^+(u)$ since f^+ is positive. Thus by induction we can show that g' has to be k-times continuously differentiable.

Now differentiation under the integral sign is allowed, and we get from theorem 1.1.1 that T_i is k-times continuously differentiable and formula (1-1-16) with Lebesgue's theorem.

Formulas for g'' and g''' are obtained via implicit differentiation of the equation $F(g(x)) = \frac{1}{2}x^2$. For a calculation of $g''(0)$ and $g'''(0)$ we refer to the next section. It is of course possible to get expressions for $g^{(4)}$, $g^{(5)}$ etc., but since they are so long that we could not figure out what to do with them they will not be mentioned. ∎

1.2 Bifurcation points and bifurcation directions

Usually, when one draws bifurcation diagrams one puts the λ on the horizontal axis and some parameter which stands for u on the vertical axis. If we take $p = u'(0)$ for this parameter then the graphs of the time maps T_i will be the solution branches, axes exchanged.

The graphs of the T_i can never intersect since it always takes a positive time for u to get from one zero to the next. Translating this into the language of bifurcation theory we have already shown that secondary bifurcation cannot occur for problem (1-0-5).

If $pT_i(p)$ is increasing (decreasing) on some interval then this corresponds to a part of the solution branch turning to the right (left). (It is always assumed that curves are run through with increasing absolute value of p.) If we look at (1-0-5) as the stationary equation of a time dependent reaction diffusion problem (t then really is a space variable), then a part of the first solution branch which is turning to the right consists of stable solutions, and a branch part turning to the left consists of unstable ones (see chapter 4, section 1).

If p goes to 0 then the corresponding u goes to 0, and the values $T_i(0)$ will just be the corresponding λ- limit points. This means that $(T_i(0), 0)$ are the bifurcation points of (1-0-5) from the trivial solution set $u \equiv 0$, $\lambda \in \mathbb{R}^+$.

$T_i'(0)$ will give the bifurcation direction of the i-th branch and, if this is zero then $T_i''(0)$ will do the job. We call a bifurcation *supercritical*, if the branch turns to the right at the bifurcation point, and *subcritical*, if it turns to the left. This way it is possible to decide whether small solutions in the first branch are stable or unstable:

$$T'(0) > 0 \Longrightarrow \begin{cases} \text{supercritical bifurcation of positive solutions (stable)} \\ \text{subcritical bifurcation of negative solutions (unstable)} \end{cases}$$

$$T'(0) < 0 \Longrightarrow \begin{cases} \text{subcritical bifurcation of positive solutions (unstable)} \\ \text{supercritical bifurcation of negative solutions (stable)} \end{cases}$$

$$T'(0) = 0, \ T''(0) > 0 \Longrightarrow \text{onesided supercritical bifurcation (stable)}$$

$$T'(0) = 0, \ T''(0) < 0 \Longrightarrow \text{onesided subcritical bifurcation (unstable)}$$

In the following proposition we will calculate $T_i^{(n)}(0)$ for $n = 0, 1, 2$ in terms of $f^{(j)}(0)$, $j = 0, 1, 2, 3$. The corresponding results on bifurcation points and bifurcation directions are all well known, but this gives us the opportunity to show how naturally all this comes out of the time map formula.

1.2.1 PROPOSITION. *Let (1-0-1) hold for f and assume that f^+ is k-times continuously differentiable.*

Then $f^{(n+1)}(0)$ exists for $0 \le n \le k$ and with $g = \Phi(f)$

$$(1\text{-}2\text{-}1) \qquad g'(0) = \frac{1}{\sqrt{f'(0)}} = \frac{1}{\sqrt{f^+(0)}} \qquad \text{for} \quad k \ge 0,$$

$$(1\text{-}2\text{-}2) \qquad g''(0) = -\frac{1}{3}\frac{f''}{(f')^2}(0) = -\frac{2(f^+)'}{3(f^+)^2}(0) \quad \text{for} \quad k \ge 1,$$

$$(1\text{-}2\text{-}3) \qquad
\begin{aligned}
g'''(0) &= \frac{1}{4(f'(0))^{7/2}}\left(\frac{5}{3}(f'')^2 - f'f'''\right)(0) \\
&= \frac{1}{4(f^+(0))^{7/2}}\left(\frac{5}{3}4((f^+)'(0))^2 - 3f^+(0)(f^+)''(0)\right).
\end{aligned}$$

$T_i^{(n)}(0)$ *for $0 \le n \le k$ is given by*

$$(1\text{-}2\text{-}4) \qquad T_i^{(n)}(0) = g^{(n+1)}(0)\int_0^{i\pi} \sin^n(\theta)\, d\theta.$$

For $n = 0, 1, 2$ we have the formulas

$$(1\text{-}2\text{-}5) \qquad T_i(0) = \frac{i\pi}{\sqrt{f'(0)}},$$

$$(1\text{-}2\text{-}6) \qquad T_i'(0) = \begin{cases} 0 & \text{if } i \text{ is even} \\ -\dfrac{2}{3}\dfrac{f''}{(f')^2}(0) & \text{if } i \text{ is odd,} \end{cases}$$

$$(1\text{-}2\text{-}7) \qquad T_i''(0) = \frac{i\pi}{2}\frac{1}{4(f'(0))^{\frac{7}{2}}}\left(\frac{5}{3}(f'')^2 - f'f'''\right)(0).$$

PROOF: The *k*-th derivative of *f* is

$$(1\text{-}2\text{-}8) \qquad f^{(k)}(u) = uf^{+(k)}(u) + kf^{+(k-1)}(u),$$

so $f^{(k+1)}(0)$ exists with

(1-2-9) $$f^{(k+1)}(0) = (k+1)f^{+(k)}(0).$$

Then (1-2-1) holds because of (1-1-6).

For (1-2-2) we have to calculate the limit of $g''(x)$ as $x \to 0$, i.e.,by (1-1-14) the limit as $u \to 0$ of

(1-2-10) $$\frac{f^2 - 2Ff'}{f^3}(u) = \frac{1}{(f^+)^3(u)}\frac{(f^2 - 2Ff')(u)}{u^3}.$$

Now using (1-2-8) (and omitting arguments):

$$\frac{f^2 - 2Ff'}{u^3} = f^+\frac{u^2 f^+ - 2F}{u^3} - (f^+)'\frac{2F}{u^2}.$$

We have already calculated that

(1-2-11) $$\lim_{u \to 0} \frac{2F(u)}{u^2} = f^+(0).$$

For the rest we use de l'Hôpital's theorem to get

$$\lim_{u \to 0}\frac{u^2 f^+ - 2F}{u^3} = \lim_{u \to 0}\frac{2u f^+ + u^2(f^+)' - 2u f^+}{3u^2}$$
$$= \frac{1}{3}(f^+)'(0).$$

Thus

(1-2-12) $$\lim_{u \to 0}\frac{f^2 - 2Ff'}{u^3} = -\frac{2}{3}(f^+)'(0)f^+(0).$$

Herefrom and from (1-2-9) the result (1-2-2) follows.

(It should be noted that these calculations and the next ones become much shorter if we just assume that f is C^{k+1} with $f'(0) > 0$. But since we have chosen to draw the assumptions a little further —)

Next we are going to prove (1-2-3).

From (1-1-15) we get that

$$g'''(x) = -g'(x)\frac{h(u)}{f^4(u)} \quad , \quad u = g(x)$$

with

$$h = 3f'(f^2 - 2Ff') + 2Fff''$$
$$= 3f'(f^2 - 2Ff') + 4Ff(f^+)' + 2Ffu(f^+)''.$$

The last term we can handle:

$$\frac{2uFf(f^+)''}{f^4} = \frac{(f^+)''}{(f^+)^3}\frac{2F}{u^2}$$
$$\rightarrow \frac{(f^+)''(0)}{(f^+(0))^2} = \frac{1}{3}\frac{f'''}{(f')^2}(0) \quad \text{as} \quad u \rightarrow 0.$$

The rest of h we denote by h_1 :

$$h_1 := 3f'(f^2 - 2Ff') + 4Ff(f^+)',$$

so

$$h_1' = 3f''(f^2 - 2Ff') - 4Ff'f'' + 2f^2f''$$
$$\quad - 2uf^2(f^+)'' - 2uFf'(f^+)'' + 4Ff(f^+)''.$$

With the substitution $2(f^+)' = f'' - u(f^+)''$

$$h_1' = 3f''(f^2 - 2Ff') - 4Ff'f'' + 2f^2f''$$
$$\quad - 2uf^2(f^+)'' - 2uFf'(f^+)'' + 4Ff(f^+)''$$
$$= 5f''(f^2 - 2Ff') + (f^+)''(4Ff - 2uf^2 - 2uFf').$$

Then

$$\lim_{u\to 0}\frac{h_1(u)}{f^4(u)} = \lim_{u\to 0}\frac{1}{(f^+(u))^4}\frac{h_1(u)}{u^4} = \frac{1}{4(f^+(0))^4}\lim_{u\to 0}\frac{h_1'(u)}{u^3},$$

and, using (1-2-11), (1-2-12):

$$\frac{h_1'}{u^3} = 5f''\frac{f^2 - 2Ff'}{u^3} + (f^+)''\left(\frac{4F}{u^2}f^+ - 2(f^+)^2 - \frac{2F}{u^2}f'\right)$$
$$\rightarrow 5f''(0)[-\frac{1}{3}(f''f')(0)] + (f^+)''(0)[2(f')^2(0) - 2(f')^2(0) - (f')^2(0)]$$
$$= -\frac{1}{3}f'(0)[5(f'')^2 + f'f''''](0).$$

Thus, putting things together:

$$\lim_{u\to 0}\frac{h(u)}{f^4(u)} = -\frac{1}{4(f'(0))^3}\left(\frac{5}{3}(f'')^2 - f'f'''\right)(0).$$

(1-2-3) then follows herefrom and from (1-2-1).

The formula for $T_i^{(n)}$ follows from (1-1-16) and Lebesgue's theorem.

(1-2-5) – (1-2-7) follow herefrom and from (1-2-1) – (1-2-3). ∎

1.3 Halfbranches without turns

In this section we will use the time map formula of theorem 1.1.1 in order to find conditions on f under which the solution branches of (1-0-5) have no turns other than at $u = 0$. We will this way recover results of Chafee, Infante in [5] and of Opial in [22].

Such branches will occur if $pT_i'(p) > 0$ for all p or if $pT_i'(p) < 0$ for all p. The corresponding boundary value problem will then have either no solution or at most two solutions with a given number of sign changes, depending on whether or not λ lies in the range of T_i. Moreover the first branch will either entirely consist of stable solutions or of unstable ones.

Looking at (1-1-16) we see that certainly $pT_i'(p) > (<)0$ if $xg''(x) > (<)0$ for all $0 \neq x \in]b^-, b^+[$, or, since x and $f(u)$ for $u = g(x)$ have the same sign, if

$$(1\text{-}3\text{-}1) \qquad (f^2 - 2Ff')(u) > 0 \quad \text{for} \quad 0 \neq u \in]a^-, a^+[,$$

or

$$(1\text{-}3\text{-}2) \qquad (f^2 - 2Ff')(u) < 0 \quad \text{for} \quad 0 \neq u \in]a^-, a^+[,$$

because of (1-1-14). This condition does not seem to be a very practical one, since F is mostly not available for a given f. But

$$\frac{d}{du}(f^2 - 2Ff')(u) = -2Ff''(u),$$

so (1-3-1) will hold if

$$(1\text{-}3\text{-}3) \qquad \qquad uf''(u) < 0,$$

and (1-3-2) if

$$(1\text{-}3\text{-}4) \qquad \qquad uf''(u) > 0.$$

Actually we need (1-3-3) only for those u for which $f'(u) > 0$ because for the other ones (1-3-1) is automatically satisfied. (1-3-3) is just the condition of Chafee and Infante in [5]. So we just proved

1.3.1 THEOREM (CHAFEE, INFANTE). *Let (1-0-1) hold for f and let f be twice continuously differentiable on $]a^-, a^+[$ with*

(1-3-5) $$uf''(u) < 0 \quad \text{for all } u \neq 0 \text{ with } f'(u) > 0.$$

Then

(1-3-6) $$pT_i'(p) > 0 \quad \text{for all } 0 \neq p \in D(T_i) \ , \ i = 1, 2, \dots .$$

If

(1-3-7) $$uf''(u) > 0 \quad \text{for all } u \neq 0$$

then

(1-3-8) $$pT_i'(p) < 0 \quad \text{for all } 0 \neq p \in D(T_i) \ , \ i = 1, 2, \dots .$$

The content of this theorem together with (1-2-5) is in shortcut notation that T_i "behaves like $t_i(u) := i\pi/\sqrt{f'(u)}$ on the interval about 0 where f' is positive", in the sense that $T_i(0) = t_i(0)$ and that $pT_i'(p)$ is positive (negative) if $ut_i'(u)$ is.

There is a more general statement than this one in the paper [22] by Opial, namely that T_i behaves like $t_i(u) := i\pi u/f(u)$. Though in [22] this has been proved for the period map rather than the Dirichlet-time-map, the same method of proof also works for our situation. The condition comes out naturally if one introduces polar coordinates in the phase plane. For our time map representation the proof has to do a little detour since $ut_i'(u) > (<)0$ does not imply that $xg''(x) > (<)0$. In our case the result follows by an integral transformation similar to the one used by Smoller and Wasserman in [31].

1.3.2 THEOREM (OPIAL). *Let (1-0-1) hold for f and assume that f^+ is continuously differentiable with*

(1-3-9) $$u\frac{d}{du}\frac{f(u)}{u} = u(f^+)'(u) < (>)0 \quad \text{for} \quad u \neq 0.$$

Then

(1-3-10) $$pT_i'(p) > (<)0 \quad \text{for} \quad 0 \neq p \in D(T_i).$$

PROOF: In formula (1-1-12) we want to make the change of variables

$$(1\text{-}3\text{-}11) \qquad\qquad g(p\sin\theta) = g(p)y,$$

so we first have to split the domain of integration into ones on which $\theta \mapsto \sin\theta$ is invertible. We can write $T_i(p)$ as a sum consisting of terms

$$\int_{2j\pi}^{(2j+1)\pi} g'(p\sin\theta)\, d\theta = 2\int_0^{\pi/2} g'(p\sin\theta)\, d\theta,$$

and of terms

$$\int_{(2j-1)\pi}^{2j\pi} g'(p\sin\theta)\, d\theta = 2\int_0^{\pi/2} g'(-p\sin\theta)\, d\theta.$$

So if we show that

$$(1\text{-}3\text{-}12) \qquad pS'(p) > (<)0 \quad\text{with}\quad S(p) := \int_0^{\pi/2} g'(p\sin\theta)\, d\theta,$$

then we are done, since $p\, d/dp\, S(-p) = (-p)S'(-p)$.

If we now perform the change of variables (1-3-11) on $S(p)$ then $g'(p\sin\theta)(p\cos\theta)\, d\theta = g(p)\, dy$, and the θ-interval $[0, \pi/2]$ is mapped to the y-interval $[0, 1]$. The function $y \mapsto (\cos\theta)^{-1}$ is integrable since it is bounded by some constant times $(1-y)^{-1/2}$:

$$\lim_{y\to 1} \frac{g(p)(1-y)}{\cos^2\theta} = \lim_{\theta\to\pi/2} \frac{g(p) - g(p\sin\theta)}{\cos^2\theta} = \frac{1}{2}pg'(p).$$

Thus we have

$$(1\text{-}3\text{-}13) \qquad\qquad S(p) = \int_0^1 \frac{g(p)}{p\cos\theta}\, dy,$$

where θ is an implicit function of y and p via (1-3-11). Implicit differentiation of (1-3-11) leads to

$$\frac{\partial\theta}{\partial p} = \frac{g'(p)g(p\sin\theta)}{g(p)g'(p\sin\theta)(p\cos\theta)} - \frac{\sin\theta}{p\cos\theta}.$$

With this and some effort the derivative of the integrand of (1-3-13) with respect to p is

$$\frac{g'(p)}{p^3\cos^3\theta}\left[p^2 - x^2 - \frac{pg(p)}{g'(p)} + \frac{xg(x)}{g'(x)}\right] \quad\text{with}\quad x = p\sin\theta.$$

With

(1-3-14)
$$h(x) := x^2 - \frac{xg(x)}{g'(x)}$$

this can be written as

$$\frac{g'(p)}{p^3 \cos \theta} \frac{h(p) - h(p \sin \theta)}{\cos^2 \theta}.$$

Since the righthandside factor of this term has a limit as $\theta \to \pi/2$ the whole function is integrable with respect to y and we get by Lebesgue's theorem that

$$pS'(p) = \frac{g'(p)}{p^2} \int_0^1 \frac{h(p) - h(p \sin \theta)}{\cos^3 \theta} \, dy.$$

This will be positive if $h'(x) > 0$ for $x > 0$, $h'(x) < 0$ for $x < 0$, and negative if $xh'(x) < 0$ for $x \neq 0$.

Looking at (1-3-14) in terms of $u = g(x)$ we have

$$h(x) = 2F(u) - uf(u) =: h_1(u) \quad \text{with} \quad u = g(x).$$

Now u and x have the same sign and $du/dx = g'(x)$ is positive. Thus $xh'(x) > (<)0$ iff $uh_1'(u) > (<)0$, and we have

$$pS'(p) > (<)0 \quad \text{for} \quad p \neq 0 \quad \text{if} \quad uh_1'(u) > (<)0 \quad \text{for} \quad u \neq 0.$$

The rest is simple:

$$h_1'(u) = f(u) - uf'(u) = -u^2 \frac{d}{du} \frac{f(u)}{u},$$

so (1-3-9) implies $uh_1'(u) > (<)0$. With this we get (1-3-12) and the theorem's assertion. ∎

1.3.3 EXAMPLES AND REMARKS.

(i) Since $d/du \, (uf'(u) - f(u)) = uf''(u)$, theorem 1.3.1 actually follows from theorem 1.3.2. The conditions of both theorems necessarily imply that $f''(0) = 0$.

(ii) Condition (1-3-5) can be easily verified for all functions f which are polynomials of degree ≥ 3 without complex zeroes and for which $f(0) = 0$,

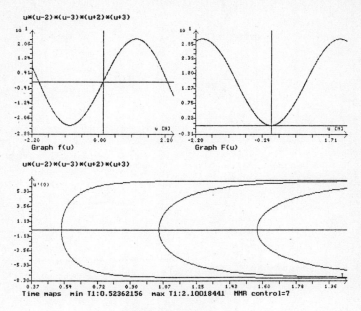

Figure 1.3.1

$f'(0) > 0$ and $f''(0) = 0$, special examples are all odd polynomials having no complex roots and for which $f(0) = 0$, $f'(0) > 0$:

a^- resp. a^+ has to be the first zero of f which is < 0 resp. > 0. (They exist since f has no complex zeroes) In each of the intervals $]a^-, 0[$ and $]0, a^+[$ the derivative f' has a unique zero, since zeroes of f and f' must alternate. The subset of $]a^-, a^+[$ on which f' is positive contains 0 and is an interval. On this interval f'' has a unique zero which is 0 by assumption. But then necessarily $uf''(u) < 0$ for all $u \neq 0$ in the interval on which f' is positive.

Figure 1.3.1 shows the computer plot of the first three time maps of such a polynomial. The T-axis is taken to be the horizontal one, so the third picture shows the bifurcation diagram of the corresponding problem (1-0-5). For information on the underlying numerics see the appendix.

In contrast to this, figure 1.3.2 shows the time maps of a polynomial for which $f''(0) \neq 0$. Here the turning point of the odd-numbered branches is different from $p = 0$, but still no other turn occurs. This way one half of the first

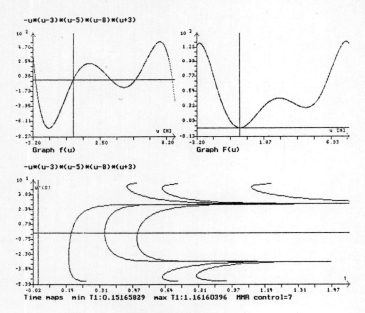

Figure 1.3.2

branch starts out unstable near the bifurcation point but recovers stability for larger absolute values of p. That this is always the case for such functions will be shown in sections 4, 5 of this chapter. The figure also shows some of the non bifurcating branches, those too seem to have but a single turning point. What can be proved of this conjecture will be discussed in chapter 3.

Using theorem 1.3.2 the class of functions with $pT'(p) > 0$ can be shown to additionally include all functions

$$f(u) = e^{h(u)}p(u)$$

p a polynomial without complex roots, $p(0) = 0$, $p'(0) > 0$, $p''(0) = 0$, $p'' \not\equiv 0$ and $uh'(u) \leq 0$:

We have seen above that $up''(u) < 0$ on the maximal interval about 0 where $p' > 0$. Thus $u \, d/du \, (p(u)/u) < 0$ on this interval. Certainly $f(0) = 0$ and $f'(0) > 0$. On the maximal interval about 0 where $f' > 0$ we have $h'p < 0$, so on this interval $0 < f' = e^h h'p + e^h p' < e^h p'$, hence $p' > 0$ on this interval. With this

$$u\frac{d}{du}\ln(\frac{f(u)}{u}) = uh'(u) + \frac{d}{du}\ln(\frac{p(u)}{u}) < 0,$$

sin(u)/cos(u)

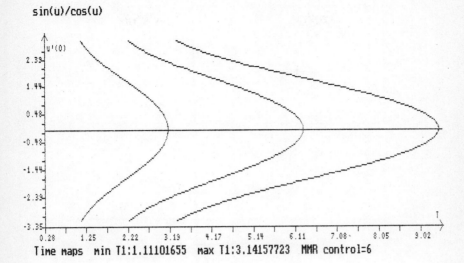

Time maps min T1:1.11101655 max T1:3.14157723 MMR control=6

Figure 1.3.3

thus $d/du\, f(u)/u < 0$ on the interval where $f' > 0$.

(iii) There are of course many examples of functions f satisfying (1-3-7). For one of those, $f(u) = \tan u$, figure 1.3.3 shows the time maps. In this example it looks like $T_i(p)$ goes to 0 as $p \to +, -\infty$. See section 1.6 for how asymptotic properties of time maps like this one can be proved.

(iv) In a way the other extreme to example (ii) are all polynomials f having only purely imaginary zeroes except for $f(0) = 0$, $f'(0) > 0$. Those f have a representation

$$f(u) = \alpha u \prod_{i=1}^{n} (u^2 + a_i^2)$$

with $\alpha > 0$, $n \geq 2$ and $a_i \neq 0$. Hereby it can easily be seen that $uf''(u) > 0$ for $u \neq 0$, so $pT_i'(p)$ is negative for those functions.

More examples for functions with (1-3-7) are all inverse functions of nonlinearities f satisfying (1-3-5), inverted on the interval where f' is positive (like arcsin or inverse funcions of odd polynomials).

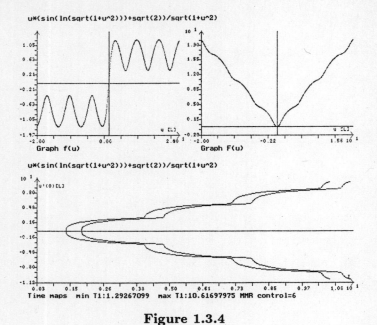

Figure 1.3.4

(v) For nonlinearities which satisfy (1-3-9) but not (1-3-5) or (1-3-7) we have to look for sublinear ((1-3-9) with ' < ') and superlinear ((1-3-9) with ' > ') functions for which f'' changes sign more than once on the set where f' is positive. Here is an example of a sublinear function f with $uf(u) > 0$ which is even oscillatory with a fixed amplitude, but of course with a frequency going to 0 as $u \to +, -\infty$: Let

$$f(u) := u\frac{\sin(\ln s) + \alpha}{s} \quad \text{with} \quad s := \sqrt{1 + u^2} \quad \text{and} \quad \alpha > \sqrt{2}.$$

Then

$$u\frac{d}{du}\frac{f(u)}{u} = \frac{\cos(\ln s) - \sin(\ln s) - \alpha}{s^2}\frac{u^2}{s},$$

which is negative for $\alpha > \sqrt{2}$. Figure 1.3.4 shows the time map picture of f with $\alpha = \sqrt{2}$. In this case the assertion of theorem 1.3.2 is also true since it does not matter if the condition is violated on a set of measure zero.

If f is a superlinear function then f' has to be positive, but there can be arbitrarily many sign changes of f'' as is shown by the example

$$f(u) = u(\cos(\alpha(s - 1)) + \gamma s), \quad s = \sqrt{1 + u^2}, \quad \gamma > |\alpha|.$$

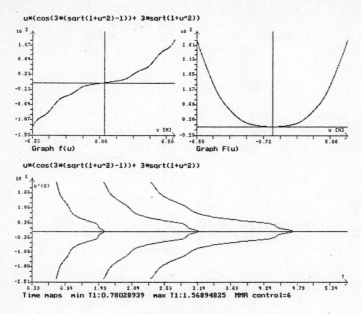

Figure 1.3.5

(see figure 1.3.5). ∎

If f is sub(super)linear only on a halfinterval $]0, a^+[$ or $]a^-, 0[$ then we get that $S'(p) > (<)0$ only for $p > 0$ or $p < 0$ (see proof of theorem 1.3.2). This means that we still get information on the branch of positive or negative solutions, but not on the higher ones:

1.3.4 PROPOSITION. *Let f satisfy (1-0-1) and let f^+ be continuously differentiable.*

If

(1-3-15)
$$\frac{d}{du}\frac{f(u)}{u} = (f^+)'(u) < (>)0 \quad on \quad]0, a^+[$$

then

(1-3-16)
$$T'(p) > (<)0 \quad on \quad]0, b^+[.$$

If

(1-3-17)
$$\frac{d}{du}\frac{f(u)}{u} = (f^+)'(u) < (>)0 \quad on \quad]a^-, 0[$$

then

(1-3-18) $T'(p) > (<)0 \quad \text{on} \quad]b^-, 0[.$

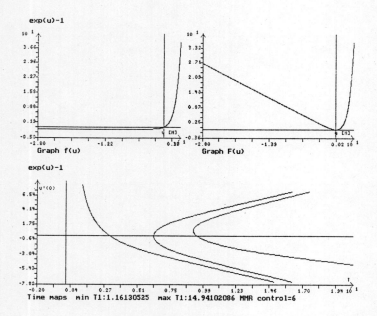

Figure 1.3.6

1.3.5 EXAMPLE.

Let

$$f(u) = e^u - 1.$$

Then f satisfies the assumptions of proposition 1.3.4 with

$$\frac{d}{du}\frac{f(u)}{u} > 0 \quad \text{on} \quad]-\infty, \infty[\setminus\{0\}.$$

Hence $T'(p)$ is negative on $D(T) =]-\infty, \infty[$. This means that negative solutions are stable and positive solutions are unstable. The opposite would of course hold for $-f(-u) = 1 - e^{-u}$. The bifurcation picture for the first three branches is shown in figure 1.3.6. Higher branches have but a single turn. This can be shown by the results of the next section.

1.4 Branches with at most one turn

In this parapraph we will give some conditions on the nonlinearity f under which time maps have at most one critical point, but not necessarily at $p = 0$. For the first branch, given by $T(p)$, this means that along each halfbranch the stability of solutions can change at most once, and that there is at least one halfbranch which does not have any change of stability.

It has been proved by Smoller and Wasserman in [31] that the time maps of

$$f(u) = -\gamma u(u - \alpha)(u - \beta) \quad \text{with } \gamma > 0,\, \alpha < 0 < \beta$$

behave like this.

By the results of this section and the next one we will be able to generalize this to all polynomial f without complex roots which satisfy condition (1-0-1) on some interval $]a^-, a^+[$.

Certainly the conditions $T_i''(p) > (<)0$ will imply such behaviour of T_i, and with proposition (1.1.2) this again follows from $g'''(x) > (<)0$. Thus, using formula (1-1-15) for g''', we get a first condition for branches without turns, namely

(1-4-1) $(3f'(f^2 - 2Ff') + 2Fff'')(u) < (>)0 \quad \text{for all} \quad u \in]a^-, a^+[\setminus \{0\}.$

There is a rare chance to be able to verify this condition in examples, so we look for a better one. It will be shown in the next theorem that (1-4-1) follows if expressions similar to the Schwarzian derivative have a single sign. We want to give names to nonlinearities with those properties, and therefor make the follwing definitions:

1.4.1 DEFINITION. *Let I be some interval and let $f : I \to \mathbb{R}$ be three times continuously differentiable. Then*

(i) *f is called an **A-function** on I iff*

(1-4-2) $$f'f''' - \frac{5}{3}(f'')^2 < 0 \quad \text{on } I.$$

(ii) *f is called a **B-function** on I iff*

(1-4-3) $$ff'' - 3(f')^2 \leq 0 \quad \text{on } I.$$

(iii) *f is called an **A-B-function** on I iff*

a) f' has only simple zeroes on I,

b) f is an A-function on any subinterval of I on which f' is positive,

c) f is a B-function on any subinterval of I on which f' is negative.

(iv) f is called a **C-function** on I iff

(1-4-4) $$f'f''' - \frac{5}{3}(f'')^2 > 0 \quad \text{on } I.$$

Classes of A-, B-, A-B- and C-functions will be discussed in the next section. Here is the main theorem of this paragraph:

1.4.2 Theorem. *Let (1-0-1) hold for f.*

(i) *If f is an A-B-function on $]a^-, a^+[$ then for all $i = 1, 2, \dots$*

(1-4-5) $$T_i''(p) > 0 \quad \text{for all } p \in D(T_i).$$

(ii) *If f is a C-function on $]a^-, a^+[$ then for all $i = 1, 2 \dots$*

(1-4-6) $$T_i''(p) < 0 \quad \text{for all } p \in D(T_i).$$

Proof: For (1-4-5) it suffices to show that g''' is positive on $]b^-, b^+[$ if f is an A-B-function.

First we show that an A-B-function can have at most one critical point on each of the intervals $]a^-, 0[$ and $]0, a^+[$:

If $f'(u) = 0$ then $f''(u) \neq 0$ by assumption. So u is the limit point of some interval where f' is negative, i.e., on this interval $ff'' - 3(f')^2$ is nonpositive. This means that $f(u)f''(u) \leq 0$, so in fact $f(u)f''(u)$ has to be negative if $f'(u) = 0$. Thus any critical point of f in $]0, a^+[$ is a strict maximum and any critical point of f in $]a^-, 0[$ is a strict minimum.

So let f' be positive on the interval $]c^-, c^+[$ containing 0, and negative on $]a^-, c^-[\cup]c^+, a^+[$. Since f is an A-B-function

(1-4-7) $$f'f''' - \frac{5}{3}(f'')^2 < 0 \quad \text{on }]c^-, c^+[.$$

Herefrom we already get $g'''(0) > 0$ by formula (1-2-3).

If we define $h(u)$ via

$$h := -3f'(f^2 - 2Ff') - 2Fff''$$

then (see (1-1-15)) $g''' > 0$ on $]b^-, b^+[$ iff $h > 0$ on $]a^-, 0[\cup]0, a^+[$.

h and h' look very similar:

$$h = -3f'(f^2 - 2Ff') - 2Fff''$$
$$h' = -5f''(f^2 - 2Ff') - 2Fff''',$$

so we get the formula

(1-4-8) $$f'h' = \frac{5}{3}f''h + 2Ff(\frac{5}{3}(f'')^2 - f'f'''),$$

which we can use to show that h cannot have a zero in $]c^-, 0[\cup]0, c^+[$:

Since $g'''(0) > 0$ we already have that $h(u) > 0$ for $0 \neq u$ near 0. If we now assume that h has a first zero u_0 in $]0, c^+[$ then from (1-4-7) and (1-4-8) it follows that $h'(u_0) > 0$, a contradiction. In the same way one shows that h cannot have a first zero u_0 in $]c^-, 0[$ (note that $f(u_0) < 0$).

if $c^+(c^-) \in]a^-, a^+[$ then $h(c^{+(-)})$ is positive since $f'(c^{+(-)}) = 0$ and $ff''(c^{+(-)})$ 0.

On the rest of $]a^-, a^+[$, where f' is negative, the positivity of h follows from (1-4-3) if we write h as

$$h = 2F(3(f')^2 - ff'') - 3f^2f'.$$

In order to show that $g''' < 0$ if f is a C-function we first note that f' is always positive on $]a^-, a^+[$ for a C-function f:

Let $]c^-, c^+[$ be the maximal interval about 0 on which f' is positive. Then on $]c^-, c^+[$

$$\frac{d^2}{du^2}(f')^{-2/3} = \frac{d}{du}(-\frac{2}{3}(f')^{-5/3}f'')$$
$$= -\frac{2}{3}(f')^{-8/3}(\frac{5}{3}(f'')^2 - f'f''') < 0$$

by assumption. But then $(f')^{-2/3}$ has to be bounded on $]c^-, c^+[$, therefore f' cannot be zero in either c^- or c^+. So $c^- = a^-$ and $c^+ = a^+$.

We can then proceed in analogy to the first part and show that $g'''(0) < 0$ and that h cannot have a first zero in either of $]0, a^+[$ or $]a^-, 0[$. ∎

Examples and figures can be found in the next section.

1.5 About A-, B-, A-B- and C-functions

1.5.1 FIRST EXAMPLES.

For some functions the A-, B- or C-property is trivial:

(i) $f(u) = e^{\alpha u} - \beta$, $\beta \in \mathbb{R}$, $\alpha > 0$ is an A-function with $f' > 0$, and therefore an A-B-function on $\mathbb{R}$. Theorem 1.4.2 can thus be applied to prove the bifurcation picture of figure 1.3.6.

(ii) $f(u) = \sin(\alpha u + \beta) + \gamma$, $\alpha \neq 0$, $|\gamma| < 1$, $\beta \in \mathbb{R}$ is both an A- and a B-function on $\mathbb{R}$ and therefore an A-B-function. If we then take β such that $\alpha \cos \beta > 0$, and $\gamma = -\cos \beta$ then we can apply theorem 1.4.2 to show that all T_i'' are positive.

(iii) Any function which is three times continuously differentiable on some interval I with $f''' < 0$ and $f' > 0$ on I is an A-B-function. An example is u^β on $\mathbb{R}^+$ with $1 < \beta < 2$.

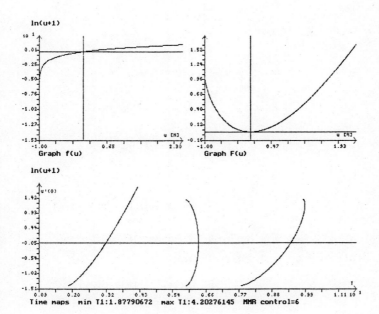

Figure 1.5.1

(iv) $f(u) = \ln(\alpha u + \beta)$ with $\alpha > 0$ is a C-function on $]-\beta/\alpha, \infty[$. Figure 1.5.1 shows the time maps of f with $\beta = 1$. The behaviour of the first branch already follows from proposition 1.3.4 since $f'' < 0$. ∎

Let us now proceed to the more general and interesting function classes:

1.5.2 PROPOSITION. *Let* $f : \mathbb{R} \to \mathbb{R}$ *be a polynomial of degree* $n \geq 2$ *with all zeroes of* f *being real.*

Then f *is a B-function on* $\mathbb{R}$, *an A-function on all intervals where* $f' \neq 0$ *and an A-B-function on all intervals which do not contain any multiple zeroes of* f.

PROOF: Let us first show that f is a B-function:

On intervals where $f \neq 0$ we have

(1-5-1)
$$(\ln|f|)'' = \frac{ff'' - (f')^2}{(f)^2}.$$

Thus for f to be a B-function we only have to show that $(\ln|f|)''$ is negative on such intervals.

Since all zeroes of f are real

$$f(u) = \alpha \prod_{i=1}^{n}(u - a_i) \quad \text{with } a_i \in \mathbb{R}, \alpha \neq 0.$$

Thus

$$\ln|f| = \ln\alpha + \sum_{i=1}^{n}\ln|u - a_i|,$$

and

$$(\ln|f|)'' = -\sum_{i=1}^{n}\frac{1}{(u - a_i)^2} < 0.$$

For f to be an A-function on intervals where $f' \neq 0$ it suffices to show that $(\ln|f'|)''$ is negative on such intervals, since

(1-5-2)
$$(\ln|f'|)'' = \frac{f'f''' - (f'')^2}{(f')^2}.$$

Now f' too can only have real zeroes since any root of f which is of order $k > 1$ corresponds to a root of f' of order $k - 1$, between any two distinct zeroes of f

there has to be a zero of f', and the multiplicities of all these real roots add up to at least $n - 1$. So in the same way as before we get that $(\ln |f'|)''$ is negative.

On an interval which contains only simple zeroes of f all zeroes of f' have to be simple too, for otherwise the multiplicities of roots of f' would add up to more than $n - 1$. Thus f is an A-B-function on such intervals. ∎

So if f is some polynomial

$$f(u) = cu \prod_{i=1}^{n}(u - a_i) \quad \text{with } c \prod_{i=1}^{n}(-a_i) > 0$$

then $f(0) = 0$, $f'(0) > 0$, and f satisfies the assumptions of theorem 1.4.2 (i) on $]a^-, a^+[$, where $]0, a^+[$ is the maximal interval on which f is positive, $]a^-, 0[$ is the maximal interval on which f is negative. We thus get that $T_i'' > 0$ for the central time maps of such f. Looking at the proof of proposition 1.5.2 one notices that the same result will still be true if we allow f or f' to have complex zeroes with real parts "far enough away" from $]a^-, a^+[$ and imaginary parts "not too big", but for reasons of simplicity we will not make a more detailed statement.

Via (1-5-1) and (1-5-2) proposition 1.5.2 can immediately be generalized to

1.5.3 PROPOSITION. *Let f be three times continuously differentiable with*

(1-5-3) $(\ln |f|)'' < 0$ *on all intervals where $f \neq 0$*

and

(1-5-4) $(\ln |f'|)'' < 0$ *on all intervals where $f' \neq 0$.*

Then f is a B-function on its definition set, an A-function on any interval where $f' \neq 0$ and an A-B-function on any interval which does not contain any multiple zeroes of f.

PROOF: That f is an A- and B-function on the intervals in question immediately follows from the assumptions. For f to be an A-B-function on an interval which only contains simple zeroes of f it remains to be shown that f' too can only have simple zeroes on such intervals:

Let u be a point in such an interval with $f'(u) = 0$. Then $f(u) \neq 0$ by assumption. Hence, using (1-5-3), $ff''(u) < 0$. ∎

This proposition is useful for building up more complicated A-B-functions out of simple ones:

If $f = f_1 f_2$ and $f' = \widetilde{f_1}\widetilde{f_2}$ then (1-5-3) and (1-5-4) hold for f if (1-5-3) holds for all factors f_i, $\widetilde{f_i}$. The statement stays true of course if we only have '$\leq$' for one factor.

We can use this in order to enlarge the class of A-B-functions we got by proposition 1.5.2:

1.5.4 PROPOSITION. *Let p be a polynomial of degree $n \geq 2$ and let α and β be real numbers.*

Then

$$f(u) = p(u)e^{\alpha+\beta u}$$

satisfies assumptions (1-5-3) and (1-5-4) of propositon 1.5.3.

PROOF: Certainly (1-5-3) holds for f. If $\beta = 0$ then we just have the situation of proposition 1.5.2. In the other case

$$f'(u) = (p'(u) + \beta p(u))e^{\alpha+\beta u},$$

and $q := p' + \beta p$ is a polynomial of degree n. For (1-5-4) to hold we have to show that all zeroes of q are real:

A zero of p which has multiplicity $k > 1$ is a zero of q with multiplicity $k - 1$. Between any two distinct roots of p there has to be a zero of f', i.e. of q. Thus if we take a_1 to be the minimal zero of p and a_2 to be the maximal one then the multiplicities of zeroes of q in $[a_1, a_2]$ add up to at least $n - 1$. Now $f(u) \to 0$ as $\beta u \to -\infty$, thus there must be an additional zero of f' in $\mathbb{R} \setminus [a_1, a_2]$. Hence q cannot have any complex zeroes.

The rest follows from proposition 1.5.3. ∎

This proposition together with Hadamard's theorem ([1], pp. 206) allows us to further enlarge the class of known A-B-functions to the set of entire functions which only have real zeroes and which have a certain growth at infinity:

1.5.5 THEOREM. *Let $f : \mathbb{C} \to \mathbb{C}$ be an entire function with $f(\mathbb{R}) \subset \mathbb{R}$, let all zeroes of f be real and assume that $f' \not\equiv const$. Further assume that*

$$(1\text{-}5\text{-}5) \qquad \limsup_{r \to \infty} \frac{\ln \ln M(r)}{\ln r} \leq 1,$$

where

$$M(r) = \max_{|z|=r} |f(z)| > 1 \quad \text{for } r \text{ large.}$$

Then f is a B-function on $\mathbb{R}$, an A-function on all subintervals of $\mathbb{R}$ where $f' \neq 0$ and an A-B-function on all subintervals of $\mathbb{R}$ which do not contain any multiple zeroes of f.

PROOF: By Hadamard's theorem a function f with the above properties has a representation

$$f(z) = e^{\alpha + \beta z} \prod_{i=1}^{\infty} b_i(z - a_i)$$

with α, β, b_i, $a_i \in \mathbb{R}$.

Let $k > 0$ be the number if indices i for which b_i is nonzero. If $2 \leq k < \infty$ then we are just in the situation of proposition 1.5.4.

If $k = 1$ then β has to be nonzero by assumption and (1-5-3), (1-5-4) can easily be shown to hold on the real line.

So let us consider the case that $k = \infty$. Then on any compact subset of $\mathbb{C}$ the functions $f^{(i)}$ are the uniform limits of $f_n^{(i)}$ with

$$f_n(z) = e^{\alpha + \beta u} \prod_{i=1}^{n} b_i(z - a_i).$$

By propositon 1.5.4 $(\ln |f_n|)''(u)$ and $(\ln |f_n'|)''(u)$ are negative for all $u \in \mathbb{R}$ where they are defined, moreover they are strictly decreasing sequences. Thus (1-5-3) and (1-5-4) hold for $f|_{\mathbb{R}}$. ∎

So we know about time maps for polynomials with real zeroes and purely imaginary zeroes (see section 1.3). An interesting question is what happens in between. If we want to homotope a poynomial with real zeroes into one with complex ones we first have to move two (say) positive roots together so they form a double zero. From there we let two conjugate complex roots move out into the complex plane.

Figures 1.5.2 – 1.5.6 show what happens during this homotopy from real to imaginary zeroes. A zero of f where $f' < 0$ lies on a homoclinic orbit in the phase plane, thus any such zero coresponds to a pole of T. This also stays true if this zero becomes a double one. So if we continue the homotopy and f has a pair of complex conjugate zeroes with small imaginary part then the pole of T vanishes

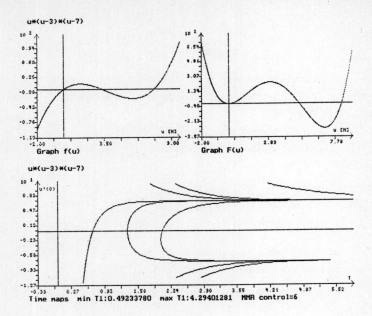

Figure 1.5.2

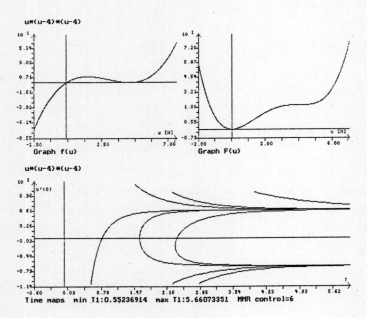

Figure 1.5.3

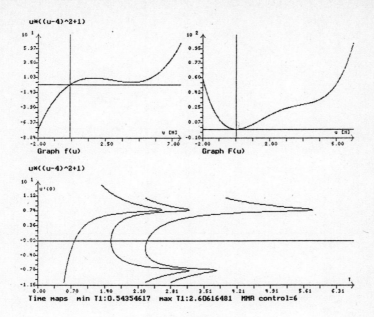

Figure 1.5.4

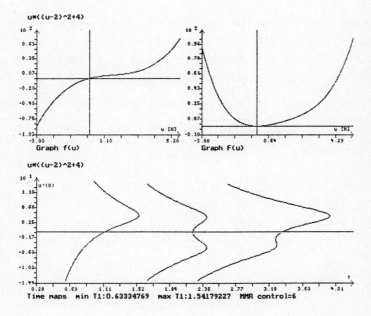

Figure 1.5.5

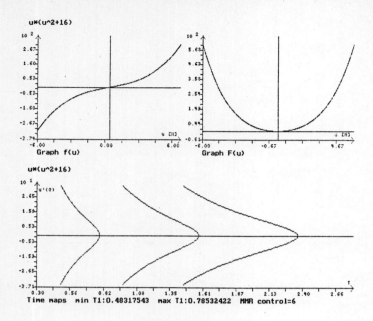

Figure 1.5.6

and T must have at least one critical point. For the cubic polynomial in the figures it looks like all halfbranches have at most one critical point. For the period map of cubics this has been shown by Chow and Sanders in [8], it looks like their method of proof would also work in the Dirichlet case.

The next proposition shows that the B-property of functions in certain cases follows from the A-property:

1.5.6 PROPOSITION. *Let* f *be an A-function on* $I = \,]a,b[\,$ *, and further assume*

$$(1\text{-}5\text{-}6) \qquad\qquad if \quad f'(u) = 0 \quad then \quad ff''(u) < 0,$$

$$(1\text{-}5\text{-}7) \quad
\begin{aligned}
&if\ f' < 0\ near\ b\ then\ f > 0\ near\ b\ and\\
&\lim_{u \to b} f(u) = 0 \quad for\ b < \infty \quad,\quad \lim_{u \to b} uf^2(u) = 0 \quad for\ b = \infty,\\
&if\ f' < 0\ near\ a\ then\ f < 0\ near\ a\ and\\
&\lim_{u \to a} f(u) = 0 \quad for\ a > -\infty \quad,\quad \lim_{u \to a} uf^2(u) = 0 \quad for\ a = -\infty.
\end{aligned}$$

Then f *is an A-B-function on* I *.*

PROOF: By assumption f has only positive maxima and negative minima. Thus the set of $u \in I$ with $f'(u) < 0$ consists of intervals $]c, d[$ with either

(1-5-8) $f|_{]c,d[} > 0$, $f'|_{]c,d[} < 0$, $f'(c) = 0$, $f(d) = 0$

(in this case d but not c can be an endpoint of I)

or

(1-5-9) $f|_{]c,d[} < 0$, $f'|_{]c,d[} < 0$, $f(c) = 0$, $f'(d) = 0$

(here c but not d can be an endpoint of I).

Without loss of generality we assume that we have an interval with (1-5-8). We have to show that
$$h := ff'' - 3(f')^2 \leq 0 \quad \text{on }]c, d[.$$

$h(c) < 0$ by assumption, and

$$h' = ff''' - 5f'f'',$$

thus

$$f'h' = \frac{5}{3}f''h - f(\frac{5}{3}f(f'')^2 - f'f'''),$$

hence

$$f'h' < \frac{5}{3}f''h$$

since f is an A-function. So for any $u \in]c, d[$ with $h(u) = 0$ it follows that $h'(u) > 0$. If we now assume that h has a zero u_0 in $]c, d[$ then h has to stay positive on $]u_0, d[$. Let us for a moment define g by $g(u) = 1/f^2(u)$. Then

$$g'' = -\frac{f''f' - 3(f')^2}{f^4} < 0 \quad \text{near } d.$$

If d is finite then this is a contradiction to $f(d) = 0$.

In the other case $d = \infty$ is the right boundary point of I and thus $g(u)/u \to \infty$ as $u \to \infty$. But from $g'' < 0$ it follows that the limit $g'(u)$ for $u \to \infty$ is finite, a contradiction.

This way we can show that f must be a B-function on any interval where f' is negative, the other properties of A-B-functions are trivial for f. ∎

This proposition gives us the opportunity to also implement the Schwarzian derivative into our system of A-B-functions. The Schwarzian derivative plays an important part in (e.g.) the theory of iterated maps on the interval (see [11]) and is
defined on the set $f' \neq 0$ by

$$(1\text{-}5\text{-}10) \qquad S f := \frac{f' f''' - \frac{3}{2}(f'')^2}{(f')^2}.$$

So it is immediate that functions with negative Schwarzian are A-functions. Now it
is a well known fact that the composition of two functions with negative Schwarzian
has again a negative Schwarzian derivative :

1.5.7 LEMMA. *Let I and J be two intervals and let $f_1 : I \to J$ and $f_2 : J \to \mathbb{R}$
be three times continuously differentiable functions.*

Then

$$(1\text{-}5\text{-}11) \qquad S(f_2 \circ f_1) = (f_1')^2 ((S f_2) \circ f_1) + S f_1$$

on the set where $(f_2 \circ f_1)' \neq 0$.

The proof of this lemma is just an exercise for using the chain rule.

Now all functions for which (1-5-4) of proposition 1.5.3 is satisfied are examples
of functions with negative Schwarzian, and many more like $e^{\alpha u + \beta} + \gamma$. A linear
fractional transformation has zero Schwarzian. With proposition 1.5.6 and lemma
1.5.7 we are now able to check whether a composition of functions with negative
Schwarzian will be an A-B-function:

1.5.8 PROPOSITION. *Let $I = \,]a, b[$ and J be intervals and assume that $f_1 :
I \to J$ and $f_2 : J \to \mathbb{R}$ are three times continuously differentiable with*

$$(1\text{-}5\text{-}12) \qquad \begin{aligned} S f_1 &\leq 0 \quad \text{and} \quad S f_2 \leq 0 \quad \text{when defined,} \\ S f_i &< 0 \quad \text{for} \quad i = 1 \quad \text{or} \quad i = 2. \end{aligned}$$

Moreover assume

$$(1\text{-}5\text{-}13) \qquad f_1 \text{ is a diffeomorphism between } I \text{ and } J \, ,$$

$$(1\text{-}5\text{-}14) \qquad \text{if} \quad f_2'(u) = 0 \quad \text{then} \quad f_2 f_2''(u) < 0 \, ,$$

if $(f_2 \circ f_1)' < 0$ *near* b *then* $f_2 \circ f_1 > 0$ *near* b *and*

(1-5-15) $\lim_{u \to b}(f_2 \circ f_1)(u) = 0$ *for* $b < \infty$,

$\lim_{u \to b} u(f_2 \circ f_1)^2(u) = 0$ *for* $b = \infty$,

if $(f_2 \circ f_1)' < 0$ *near* a *then* $f_2 \circ f_1 < 0$ *near* a *and*

(1-5-16) $\lim_{u \to a}(f_2 \circ f_1)(u) = 0$ *for* $a > -\infty$,

$\lim_{u \to a} u(f_2 \circ f_1)^2(u) = 0$ *for* $a = -\infty$.

Then $f_2 \circ f_1$ *is an A-B-function on* I.

PROOF: On subintervals of I where $(f_2 \circ f_1)' = f_1'(f_2' \circ f_1)$ is nonzero $f_2 \circ f_1$ is an A-function because of condition (1-5-12) and lemma 1.5.7.

If $(f_2 \circ f_1)'(u) = 0$ then $f_2'(f_1(u))$ has to vanish since f_1' is never zero. Thus by (1-5-14)

$$(f_2 \circ f_1)(f_2 \circ f_1)''(u) = f_2 f_2''(f_1(u)) < 0.$$

Hence $f_2 \circ f_1$ is an A-function on I and the assumptions of proposition 1.5.6 are fulfilled for $f_2 \circ f_1$. ∎

1.5.9 EXAMPLES.

The following examples are meant to illustrate the use of proposition 1.5.8:

(i) Let p be some polynomial with all roots of p being real and nonnegative, and let further be $p(1) = 0$ with $p'(1) > 0$. Consider p on the interval $J =]c^-, c^+[$ with $]1, c^+[$ being the maximal interval on which p is positive, $]c^-, 1[$ being the maximal subinterval of $]0, 1[$ on which p is negative. Then p satisfies the assumptions on f_2 in proposition 1.5.8 on I. As f_1 we can take e^u on $I =]a^-, a^+[\ = \ ]\ln c^-, \ln c^+[$.

It remains to be checked that the limit conditions in (1-5-15) and (1-5-16) are satisfied at the boundary of I. If 0 is not a zero of p then either p' is positive on $]0, 1[$ or $c^- > 0$, i.e. $a^- > -\infty$, in both cases (1-5-16) holds at $a = a^-$. For $p(0) = 0$ we have $p(e^u) = e^u q(e^u)$, q being some polynomial. So, if $c^- = 0$ in this case, we get $\lim_{u \to -\infty} u p^2(e^u) = 0$, and (1-5-16) holds at $a = a^- = -\infty$. To prove the limit conditions at $b = a^+$ is analogous. So the function

$$u \mapsto p(e^u)$$

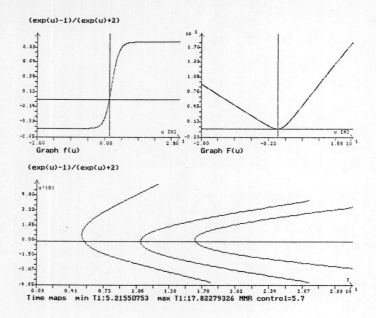

Figure 1.5.7

satisfies the assumptions of theorem 1.4.2 (i) on $]a^-, a^+[$.

(ii) A less complicated example is

$$f(u) = \frac{e^u - 1}{\alpha e^u + \beta} \quad \text{with } \alpha, \beta > 0.$$

Taking $f_2(v) = (v-1)/(\alpha v + \beta)$ and $f_1(u) = e^u$ we see that the conditions of proposition 1.5.8 are satisfied for $I = \mathbb{R}$ and $J = \mathbb{R}^+$ since $S f_2 = 0$, $S f_1 > 0$ and both f_2' and f_1' are positive. So T_i'' is always positive for f (see figure 1.5.7).

(iii) In the same way we can show that the assumptions of theorem 1.4.2 (i) hold for

$$f(u) = e^{(\alpha u + \beta)/(\gamma u + \delta)} - e^{\beta/\delta} \quad \text{with } \alpha\delta - \beta\gamma > 0, \, \delta \neq 0$$

if we take for I the subinterval of the set $u \neq \delta/\gamma$ in which 0 lies . ∎

For the last bit of this section let us discuss C-functions. It seems like there is not so much to say about these as about A-B-functions.

f being a C-function is equivalent to

$$(|f'|^{-2/3})'' < 0.$$

This way, as we have already seen, f' cannot have any zeroes on the definition set of f. Moreover, since $|f'|$ is always bounded from below, f can never be defined without singularities on the whole real line.

There is a corresponding feature of the time maps of C-functions: Since T_i'' is negative and T_i has to stay positive, time maps can never be defined for all $p \in \mathbb{R}$, and either the branch of positive solutions or the one of negative solutions has to "stop dead" right in the middle of the set $\{\lambda > 0, u'(0) \neq 0\}$. At first sight this is a contradiction to Rabinowitz's global bifurcation theorem ([24]), which it is not, of course: At the "stopping point" the problem leaves the region where it is reasonably defined, and this possibility too is part of the Rabinowitz-alternative. (See the introduction.)

Here is a general criterion for C-functions:

1.5.10 PROPOSITION. *Let $\tilde{f}$ be three times continuously differentiable on some interval J with*

$$(\ln |\tilde{f}'|)'' < 0 \quad , \quad \tilde{f}' \neq 0 \quad \text{on } J.$$

Let $I := \tilde{f}^{-1}(J)$ and let $f : I \to J$ be the inverse function of $\tilde{f}$.

Then f is a C-function on I.

PROOF: Let $u \in I$, and let $u = \tilde{f}(v)$. Then by implicit differentiation

$$f'(u) = \frac{1}{\tilde{f}'}(v) \quad f''(u) = -\frac{\tilde{f}''}{(\tilde{f}')^3}(v) \quad f'''(u) = \frac{3(\tilde{f}'')^2 - \tilde{f}'\tilde{f}'''}{(\tilde{f}')^5}(v).$$

With this

$$(3f'f''' - 5(f'')^2)(u) = -\frac{3}{(\tilde{f}')^6}\left(\tilde{f}'\tilde{f}''' - \frac{4}{3}(\tilde{f}'')^2\right)$$

$$\leq \frac{3}{(f')^4}(\ln|\tilde{f}'|)'' < 0.$$

Thus f is a C-function on I. ■

By above results on A-functions this class e.g. includes inverse functions of polynomials p for which p' has no complex zeoes.

1.6 Asymptotic behaviour of time maps

In this section we will use the time map formula of theorem 1.1.1 in order to see which is the behaviour of time maps near the boundary of their definition set. If one knows already that T has at most one critical point then information on the asymptotic behaviour can decide upon whether there has to be at least one.

With f satisfying (1-0-1) we use the time map formulas from section 1.1:

$$T(p) = 2 \int_0^{g(p)} \frac{du}{\sqrt{p^2 - 2F(u)}} mtag$$

and

(1-6-1) $$T(p) = 2 \int_0^{\pi/2} g'(p \sin \theta) \, d\theta$$

for $p \in D(T) =]b^-, b^+[\setminus \{0\}$. See section 1.1 for more details.

If $f(a^\kappa) = 0$ ($\kappa = +$ or $-$) such that $f'(a^\kappa) \leq 0$ exists then $(a^\kappa, 0)$ is a homoclinic point of the equation $u'' + f(u) = 0$ in the phase plane, the corresponding homoclinic orbit crosses the $u = 0$-axis at $(0, b^\kappa)$. Thus $T(p)$ goes to $+\infty$ as $p \to b^\kappa$. In the following proposition this assumption is generalized to f' being bounded from below near a^κ or f' decreasing "not too fast" towards $-\infty$. On the other hand, if f' is decreasing with a higher velocity then the solution $U(\cdot, b^\kappa)$ will reach its extremum a^κ in finite time, and T approaches a finite value as $p \to b^\kappa$. This fact can be used to answer some questions about "dead core problems" on the line (see example 1.6.2).

1.6.1 PROPOSITION. *Let (1-0-1) hold for* f *and let* κ *denote* $+$ *or* $-$.

(i) *Assume that*

(1-6-2) $$|a^\kappa| < \infty \quad , \quad \lim_{u \to a^\kappa} f(u) = 0 ,$$

(1-6-3) $$|f(u)| \leq c|a^\kappa - u| (\ln |a^\kappa - u|)^2$$

for u *near* a^κ *with some constant* $c > 0$.

Then $T(p) \to \infty$ *as* $p \to b^\kappa$.

(ii) *Assume that*

(1-6-4) $$|a^\kappa| < \infty \quad , \quad \lim_{u \to a^\kappa} f(u) = 0 ,$$

(1-6-5)
$$|f(u)| \geq c|a^{\kappa} - u|(-\ln|a^{\kappa} - u|)^q$$
for u near a^{κ} with constants $c > 0$, $q > 2$.

Then $T(p)$ has a positive and finite limit as $p \to b^{\kappa}$.

(iii) Let

(1-6-6)
$$|a^{\kappa}| < \infty \quad, \quad \lim_{u \to a^{\kappa}} f(u) = \kappa\infty,$$

(1-6-7)
$$f'(u) > 0 \quad \text{for } u \text{ near } a^{\kappa}.$$

If $F(a^{\kappa}) < \infty$ then $T(p)$ has a positive and finite limit as $p \to b^{\kappa}$.
If $F(a^{\kappa}) = \infty$ then $T(p) \to 0$ as $p \to b^{\kappa} = \kappa\infty$.

(iv) Assume that

(1-6-8)
$$|a^{\kappa}| = \infty \quad, \quad \lim_{u \to a^{\kappa}} \frac{f^2(u)}{2F(u)} = c^{\kappa} \quad \text{with } 0 \leq c^{\kappa} \leq \infty.$$

Then $T(p) \to \pi/\sqrt{c^{\kappa}}$ for $p \to b^{\kappa}$. (With $1/0 := +\infty$, $1/\infty := 0$.)

PROOF: Without loss of generality we only prove the results for $\kappa = +$.
Ad (i): From (1-6-1) we get

$$T(p) \geq 2 \int_0^{g(p)} \frac{du}{\sqrt{(b^+)^2 - 2F(u)}}.$$

For u close to a^+ via (1-6-4)

$$(b^+)^2 - 2F(u) = 2(F(a^+) - F(u)) = 2 \int_u^{a^+} f(v)\, dv$$

$$\leq 2c \int_u^{a^+} (a^+ - v)(\ln(a^+ - v))^2\, dv$$

$$\leq 2c \int_u^{a^+} 2(a^+ - v)(-\ln(a^+ - v))(-\ln(a^+ - v) - 1)\, dv$$

since $-\ln(a^+ - v) \geq 2$ for v close to a^+. The integral in the last term is calculated explicitly to

$$(a^+ - u)^2(-\ln(a^+ - u))^2.$$

Thus for $\delta > 0$ small enough and p close to b^+ :

$$T(p) \geq \int_{a^+-\delta}^{g(p)} \frac{du}{\sqrt{(b^+)^2 - 2F(u)}}$$

$$\geq \frac{1}{\sqrt{2c}} \int_{a^+-\delta}^{g(p)} \frac{du}{(a^+ - u)(-\ln(a^+ - u))}$$

$$= \frac{1}{\sqrt{2c}} \left[\ln(-\ln(a^+ - g(p))) - \ln(-\ln \delta) \right] \longrightarrow \infty \quad \text{as } p \to b^+$$

since $g(p) \to a^+$ as $p \to b^+$.

Ad (ii): Using Lebegue's theorem it suffices to show that

$$\int_{\pi/2-\delta}^{\pi/2} g'(b^+ \sin \theta) \, d\theta < \infty,$$

or, by the transformation $u = g(p \sin \theta)$, that

(1-6-9)
$$\int_{a^+-\delta}^{a^+} \frac{du}{\sqrt{(b^+)^2 - 2F(u)}} < \infty$$

for some $\delta > 0$ small.

With (1-6-6)

$$(b^+)^2 - 2F(u) = 2(F(a^+) - F(u)) = 2 \int_u^{a^+} f(v) \, dv$$

$$\geq 2c \int_u^{a^+} (a^+ - v)(-\ln(a^+ - v))^q \, dv$$

$$\geq c \int_u^{a^+} 2(a^+ - v)(-\ln(a^+ - v))^q - q(a^+ - v)(-\ln(a^+ - v))^{q-1} \, dv$$

$$= -c(a^+ - v)^2(-\ln(a^+ - v))^q \Big|_u^{a^+} = c(a^+ - u)^2(-\ln(a^+ - u))^q.$$

Hence

$$\int_{a^+-\delta}^{a^+} \frac{du}{\sqrt{(b^+)^2 - 2F(u)}} \leq \frac{1}{\sqrt{c}} \int_{a^+-\delta}^{a^+} \frac{du}{(a^+ - u)(-\ln(a^+ - u))^{q/2}}$$

$$= \frac{1}{\sqrt{c}} \frac{2}{2 - q}(-\ln(a^+ - u))^{(2-q)/2} \Big|_{a^+-\delta}^{a^+}$$

$$= \frac{1}{\sqrt{c}} \frac{2}{q - 2}(-\ln \delta) < \infty$$

since $2 - q < 0$. So we have shown (1-6-10) and the assertion of (ii).

Ad (iii): Let us first consider the case that $b^+ = \sqrt{2F(a^+)} < \infty$. Then

$$(g'(x))^2 = \frac{2F(u)}{f^2(u)} \quad \text{with } u = g(x).$$

(See (1-1-5).) From the assumptions it follows that $g'(x) \to 0$ as $x \to b^+$. Thus $\theta \mapsto g'(b^+ \sin \theta)$ is finite on $[0, \pi/2]$ and positive on $]0, \pi/2[$, and

$$T(p) = 2 \int_0^{\pi/2} g'(p \sin \theta) \, d\theta \longrightarrow 2 \int_0^{\pi/2} g'(b^+ \sin \theta) \, d\theta,$$

a positive and finite limit.

If $b^+ = \sqrt{2F(a^+)} = \infty$ then

$$\lim_{x \to \infty} (g'(x))^2 = \lim_{u \to a^+} \frac{2F(u)}{f^2(u)} = 0$$

since

$$F(u) = \int_0^u f(s) \, ds \le u f(u)$$

for u close to a^+ because of the assumptions.

Thus

$$\lim_{p \to \infty} g'(p \sin \theta) = 0 \quad \text{for } \theta \in \,]0, \pi/2[$$

and $g'(p \sin \theta)$ stays bounded near $\theta = 0$.

Hence with Lebegue's theorem

$$\lim_{p \to b^+} T(p) = 0.$$

Ad (iv): As a first case we again consider $b^+ = \sqrt{2F(a^+)} = \sqrt{2F(\infty)} < \infty$.

Since $\lim_{u \to \infty} f^2(u)/(2F(u))$ exists we also have that $\lim_{u \to \infty} f(u)$ exists, so $f(u) \to 0$ as $u \to \infty$. Thus

$$c^+ = 0 \quad , \quad \lim_{x \to b^+} g'(x) = \frac{1}{\sqrt{c^+}} = \infty.$$

Since $g(p) \to a^+ = \infty$ as $p \to b^+$

$$T(p) = 2 \int_0^{g(p)} \frac{du}{\sqrt{p^2 - 2F(u)}}$$

$$\ge \frac{2}{p} g(p) \longrightarrow \infty = \frac{\pi}{\sqrt{c^+}} \quad \text{as } p \to b^+.$$

Now let $b^+ = \sqrt{2F(a^+)} = \sqrt{2F(\infty)} = \infty$.

Then

$$\lim_{x\to\infty} (g'(x))^2 = \lim_{u\to a^+} \frac{2F(u)}{f^2(u)} = \frac{1}{c^+},$$

and by Lebegue's theorem

$$T(p) = 2 \int_0^{\pi/2} g'(p\sin\theta)\, d\theta \longrightarrow \frac{\pi}{\sqrt{c^+}} \quad \text{as } p \to b^+ = \infty.$$

∎

REMARKS:

(1-6-4) follows from

(1-6-10) $f'(u) \geq -c(\ln |a^\kappa - u|)^2,$

and (1-6-6) follows from

(1-6-11) $f'(u) \leq -c(-\ln |a^\kappa - u|)^q.$

The behaviour of T if $|a^\kappa| = \infty$ is also described by $f(u)/u$ or $f'(u)$:

(1-6-12)

$$\text{If } \quad 0 < \lim_{u\to a^\kappa} \frac{f(u)}{u} < \infty \quad \text{exists}$$

$$\text{then} \quad \lim_{u\to a^\kappa} \frac{f^2(u)}{2F(u)} = \lim_{u\to a^\kappa} \frac{f(u)}{u}.$$

(1-6-13)

$$\text{If } \quad 0 \leq \lim_{u\to a^\kappa} f'(u) \leq \infty \quad \text{exists}$$

$$\text{then} \quad \lim_{u\to a^\kappa} \frac{f^2(u)}{2F(u)} = \lim_{u\to a^\kappa} f'(u).$$

The proofs of these facts are trivial. ∎

With proposition (1.6.1) we are able to prove the limiting behaviour of T for all
examples so far discussed. For $f(u) = \tan u$ as one example we have $F(u) =$
$-\ln(\cos u) \to \infty$ as $u \to \pm\pi/2$. Thus via (iii) $T(p) \to 0$ as $p \to b^\kappa = \kappa\infty$. The
other calculations are left to the reader.

If the assumptions of (ii) are satisfied then T approaches a value $0 < T_0 < \infty$ as $p \to b^\kappa$. This means that the solution $U(\cdot, b^\kappa)$ of $u'' + f(u) = 0$ reaches the value a^κ in finite time $T_0/2$ (see section 1.1). But $u \equiv a^\kappa$ also is a solution to $u'' + f(u) = 0$. So we are able to solve

$$u'' + f(u) = 0$$

$$u(0) = u(\lambda) = 0$$

for any $\lambda > T_0$ by inserting a piece $u \equiv a^\kappa$ of appropriate length in between the solutions $U(\cdot, b^\kappa)$ and $U(2T_0 - \cdot, b^\kappa)$ (see figure 1.6.1).

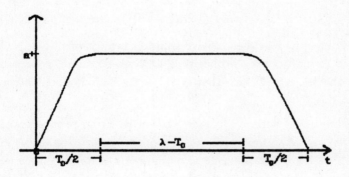

Figure 1.6.1

Here is an application:

1.6.2 EXAMPLE (A DEAD CORE PROBLEM).

The equation

(1-6-14)

$$v'' + h(v) = 0$$

$$v(0) = c = v(\lambda)$$

with $c, \lambda \in \mathbb{R}^+$ can be understood to model the stationary temperature distribution of a chemical reaction "on a line" aof length λ, with temperatures being kept at value $c > 0$ at the boundaries.

The question is whether for certain h and λ it is possible to get a "dead core" inside $]0, \lambda[$, i.e., an interval on shich some solution v vanishes identically.

As a concrete example let us take

$$h(v) = -v(1 - v + (-\ln v)^q) \quad \text{for } 0 < v < 1$$

with

$$\overset{*}{q} > 1 \quad \text{and} \quad c = 1.$$

With the change of variables $u := 1 - v$ problem (1-6-15) becomes

(1-6-15)
$$u'' + f(u) = 0$$
$$u(0) = u(\lambda) = 0$$

with

$$f(u) = (1 - u)(u + (-\ln(1 - u))^q) \quad \text{for } 0 < u < 1.$$

We are only interested in the positive solution branch of (1-6-16) , but in order to be able to formally apply proposition 1.6.1 we continue f as a differentiable function to the set $u \leq 0$ such that $f|_{\mathbb{R}_-} < 0$. This is possible since $f(0) = 0$ and $f'(0) = 1$.

So f satisfies (1-0-1) with $a^- = -\infty$, $a^+ = 1$. From proposition 1.6.1 (i), (ii) it then follows that solutions with $u \equiv 1$ on a subset of $]0, \lambda[$, λ large enough, occur for $q > 2$, for $q \leq 2$ the time to reach $u = 1$ is infinite. As a result (1-6-15) has dead core solutions for $q > 2$ whereas for $q \leq 2$ all solutions stay positive. (See also example 3.2.2.)

The asymptotic behaviour of the time maps T_i for $i > 1$ can be calculated from the following formulas which are easily derived from (1-1-13) or by observing that a solution with a given number of sign changes is made up of positive and negative solutions:

Let $p \in D(T_i) =] - \bar{b}, \bar{b}[$ with $\bar{b} = \min(b^+, -b^-)$ (see section 1.1). Then

(1-6-16) $$T_i(p) = j(T(p) + T(-p)) \quad \text{if } i = 2j \text{ is even}$$

(1-6-17) $$T_i(p) = (j + 1)T(p) + jT(-p) \quad \text{if } i = 2j + 1 \text{ is odd}.$$

The crucial point for calculating the limits of $T_i(p)$ is to observe which boundary point of $] - \bar{b}, \bar{b}[$ coincides with a boundary point of $]b^-, b^+[$.

CHAPTER II

NEUMANN PROBLEMS, PERIOD MAPS AND SEMILINEAR DIRICHLET PROBLEMS

2.1 Bifurcating Neumann branches

Nontrivial branches of the Neumann problem

(2-1-1)
$$u'' + f(u) = 0$$
$$u'(0) = u'(\lambda) = 0$$

bifurcate from any trivial branch $\{(\lambda, u) \mid \lambda \in \mathbb{R}^+ , u(t) \equiv r\}$ such that $f(r) = 0$, $f'(r) > 0$.

In these branches solutions are bounded by the first zeroes below and above r of f. So in analogy to chapter I we impose the following conditions on f:

(2-1-2)
$$f(u) = (u - r)f^+(u) \quad \text{with } a^- < r < a^+ \text{ and}$$
$$f^+ :]a^-, a^+[\to \mathbb{R}^+ \quad \text{locally Lipschitz-continuous.}$$

Thus

(2-1-3)
$$u'' + f(u) = 0$$
$$u'(0) = 0, \ u(0) = a$$

has a unique solution $t \mapsto U(t, a)$ for any $a \in]a^-, a^+[$. If $U(\cdot, a) \not\equiv const$ then $U'(\cdot, a)$ ("$'$" denoting differentiation with respect to t) has to have isolated zeroes. Then we can define the Neumann-time-map T of (2-1-3) by

(2-1-4)
$$D(T) := \{a \in]a^-, a^+[\mid U(\cdot, a) \not\equiv const, \ U'(t, a) = 0 \text{ for some } t > 0\},$$
$$T(a) := \min\{t > 0 \mid U'(t, a) = 0\} \quad \text{for } a \in D(T).$$

There is no reason for introducing T_i in analogy to T_i in order to describe solutions with one or more sign changes of u', because a solution is always symmetric in a critical point. Thus any solution $t \mapsto U(t, a)$ with $a \in D(T)$ is periodic, and

(2-1-5)
$$\text{if } a \in D(T) \text{ then}$$
$$iT(a) \quad \text{is the } i\text{-th zero of } U'(\cdot, a).$$

So we get the following relation between (2-1-1) and the Neumann-time-map:

(λ, u) is a solution of (2-1-1) with $u \not\equiv const$

iff

$u(t) = U(t, a)$ for some $a \in D(T)$

and

$\lambda = iT(a)$ for some $i \in \{1, 2, ...\}$.

The inverted graphs $(iT(a), a)$ thus give the branches of (2-1-1) projected to the $(\lambda, u(0))$-plane.

By a little observation we are able to apply the results for Dirichlet-time-maps to the Neumann case:

First of all we could as well assume that $r = 0$, for via $\tilde{u} := u - r$ we transform (2-1-3) to

(2-1-6)
$$\tilde{u}'' + \tilde{f}(\tilde{u}) = 0$$
$$\tilde{u}'(0) = 0, \ \tilde{u}(0) = a - r$$

with

$$\tilde{f}(u) := f(\tilde{u} + r).$$

So $\tilde{f}$ satisfies (1-0-1) on $]\tilde{a}^-, \tilde{a}^+[:=]a^- - r, a^+ - r[$. $\tilde{u}'$ and u' have the same zeroes, thus

(2-1-7)
$$T(a) = \tilde{T}(a - r),$$

if $\tilde{T}$ is the time map of $\tilde{f}$.

If now $\tilde{u}(t) = \tilde{U}(t, \tilde{a})$ is some nonconstant solution of (2-1-6) with initial value $\tilde{a} \in]\tilde{a}^-, \tilde{a}^+[$ (i.e. $\tilde{a} \neq 0$) then $\tilde{u}$ has to cross the level 0 between any two zeroes of $\tilde{u}'$ since $\tilde{u}''$ and thus $\tilde{f}$ has to be of opposite sign at two consecutive zeroes of $\tilde{u}'$. Let t_0 be the first zero of $\tilde{u}$, and let $p := -\tilde{u}'(t_0)$. Then $p \neq 0$ and the next zero of $\tilde{u}$ (which has to exist since $\tilde{u}$ is periodic) is given by $t_0 + T(-p)$, T being the Dirichlet-time-map of

$$u'' + \tilde{f}(u) = 0$$
$$u(0) = 0, \ u'(0) = p.$$

Then the next zero of $\tilde{u}'$ is just $t_0 + T(-p)/2$, so

$$\tilde{T}(\tilde{a}) = t_0 + T(-p)/2.$$

We can also calculate t_0:

Because of symmetry conditions

$$\tilde{u}(t_0 + T(-p)/2 + t) = \tilde{u}(t_0 + T(-p)/2 - t)$$
$$\tilde{u}'(t_0 + T(-p)/2 + t) = -\tilde{u}'(t_0 + T(-p)/2 - t).$$

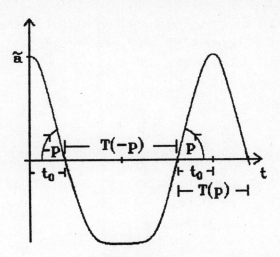

Figure 2.1.1

Thus $\tilde{u}(t_0 + T(-p)) = 0$, $\tilde{u}'(t_0 + T(-p)) = p$, and the next zero of $\tilde{u}'$ which is $> t_0 + T(-p)$ must be $2t_0 + T(-p)$. The time to go from one zero of $\tilde{u}$ to the next critical point is again half the time it takes to go to the next zero, thus

$$t_0 = (2t_0 + T(-p)) - (t_0 + T(-p)) = T(p)/2.$$

So we finally get

$$\tilde{T}(\tilde{a}) = \frac{1}{2}(T(p) + T(-p)),$$

$p = -\tilde{u}'(t_0)$ for the first zero t_0 of $\tilde{u} = \tilde{U}(\cdot, \tilde{a})$.

See figure 2.1.1 for a more illustrative proof.

It remains to establish a connection between $\tilde{a}$ and p:

Let

$$\tilde{F}(u) := \int_0^u \tilde{f}(s)\, ds\,,$$

then (2-1-6) integrates to

$$\frac{1}{2}(\tilde{u}')^2 + \tilde{F}(\tilde{u}) \equiv const = \tilde{F}(\tilde{a})\,,$$

thus for any zero t_0 of $\tilde{u}$ we have

$$\frac{1}{2}p^2 = \frac{1}{2}(\tilde{u}'(t_0))^2 = \tilde{F}(\tilde{a}).$$

If now $\tilde{a}$ is positive then $\tilde{u}'(t_0) = -p$ will be negative for the first zero of $\tilde{u}$, if $\tilde{a}$ is negative then $\tilde{u}'(t_0) = -p > 0$. So if we define

$$g(x) := \Phi(\tilde{f})(x)$$

(see (1-1-4)) then

$$\tilde{a} = g(p).$$

With $\tilde{T}(\tilde{a}) = T(\tilde{a} + r)$

$$T(g(p) + r) = \frac{1}{2}(T(p) + T(-p)) = \frac{1}{2}T_2(p)$$

(see (1-6-17)).

This way we can also calculate the definition set of T:

$T(p)$ and $T(-p)$ have to be defined, so p has to be in $]-\bar{b}, \bar{b}[$ with $\bar{b} = \min(\sqrt{2\tilde{F}(\tilde{a}^-)}, \sqrt{2\tilde{F}(\tilde{a}^+)}) = \min(\sqrt{2F(a^-)}, \sqrt{2F(a^+)})$, if we define $F(u) := \tilde{F}(u - r)$.

Thus

$$D(T) \cup \{r\} =]g(-\bar{b}) + r, g(\bar{b}) + r[$$

which is just the maximal subinterval $]\underline{a}, \bar{a}[$ of $]a^-, a^+[$ with

$$\int_{\underline{a}}^{\bar{a}} f(u)\, du = 0.$$

We collext results in the following proposition:

2.1.1 PROPOSITION. *Let (2-1-2) hold for f, and define $\tilde{f}$ by*

$$\tilde{f}(u) := f(u + r).$$

Then $\tilde{f}$ satisfies (1-0-1) on $]a^- - r, a^+ - r[$.

Define F to be the integral of f with $F(r) = 0$, and let $]\underline{a}, \bar{a}[$ be the maximal subinterval of $]a^-, a^+[$ such that $F(\bar{a}) = F(\underline{a})$.

Then

$$D(T) =]\underline{a}, \bar{a}[\setminus \{r\}.$$

Further let $b^- := \sqrt{2F(a^-)}$, $b^+ := \sqrt{2F(a^+)}$, define

$$g :]b^-, b^+[\to]a^- - r, a^+ - r[$$

by

$$g(x) = \Phi(\tilde{f})(x),$$

and let $\bar{b} := \min(b^+, -b^-)$.

Then $g :] - \bar{b}, \bar{b}[\to]\underline{a} - r, \bar{a} - r[$ is a diffeomorphism, and

(2-1-8) $$\mathcal{T}(g(p) + r) = \frac{1}{2}T_2(p) = \frac{1}{2}(T(p) + T(-p))$$

with T, T_2 being the Dirichlet-time-maps of $\tilde{f}$.

So Neumann branches are nothing but transformed even numbered Dirichlet branches. With $a = g(p) + r$

(2-1-9) $$\mathcal{T}'(a)g'(p) = \frac{1}{2}T_2'(p)$$

(2-1-10) $$\mathcal{T}''(a)(g'(p))^2 + \mathcal{T}'(a)g''(p) = \frac{1}{2}T_2''(p),$$

provided $\mathcal{T}'$, $\mathcal{T}''$ exist.

From proposition 1.2.1 we thus get the following formulas for bifurcation points and bifurcation directions if f has the appropriate regularity:

(2-1-11) $$\mathcal{T}(r) = \frac{1}{2}T_2(0) = \frac{\pi}{\sqrt{f'(r)}}$$

(2-1-12) $$\mathcal{T}'(r) = \frac{1}{2}\frac{T_2'(0)}{g'(0)} = 0$$

(2-1-13) $$\mathcal{T}''(r) = \frac{1}{2}\frac{T_2''(0)}{(g'(0))^2} = \frac{\pi}{4(f'(r))^{5/2}}(\frac{5}{3}((f'')^2 - f'f''')(r).$$

Also, since $g' > 0$, $\mathcal{T}'(a)$ and $T_2'(p)$ for $a = g(p) + r$ have the same sign as well as $(a - r)\mathcal{T}'(a)$ and $pT'(p)$ since p and $g(p)$ have the same sign. Hence the following theorem is a consequence of theorem 1.3.2:

2.1.2 THEOREM (OPIAL). *Let (2-1-2) hold for f and assume that f^+ is continuously differentiable with*

$$(2\text{-}1\text{-}14) \qquad (u - r)\frac{d}{du}\frac{f(u)}{(u-r)} = (u-r)(f^+)'(u) < (>)0 \quad \text{for} \quad u \neq r.$$

Then

$$(a - r)T'(a) > (<)0 \quad \text{for} \quad a \in]\underline{a}, \bar{a}[\setminus \{r\}.$$

If T_2'' is of one sign then it follows from (2-1-10) that T' can have at most one zero and that at this zero T'' has the sign of T_2''. Since $T'(r)$ is always zero we have that $(a - r)T'(a)$ for $a \neq r$ has the same sign as T_2''. The A-, B-, A-B- or C-properties of functions are preserved under a shift in the argument. So we can derive the following theorem from theorem 1.4.2:

2.1.3 THEOREM. *Let (2-1-2) hold for f.*

(i) *If f is an A-B-function on $]a^-, a^+[$ then*

$$(a - r)T'(a) > 0 \quad \text{for} \quad a \in]\underline{a}, \bar{a}[\setminus \{r\}.$$

(ii) *If f is a C-function on $]a^-, a^+[$ then*

$$(a - r)T'(a) < 0 \quad \text{for} \quad a \in]\underline{a}, \bar{a}[\setminus \{r\}.$$

With this theorem we can generalize the result for the Neumann problem in [28], where f is assumed to be a cubic

$$f(u) = -\delta(u - \alpha)(u - \beta)(u - \gamma) \quad , \quad \delta > 0, \, \alpha < \beta < \gamma,$$

to the case of f being any polynomial with real zeroes, r being some zero of f with $f'(r) > 0$. This result and many more follow from section 1.5.

It should be noted that in the Neumann case we cannot deduce any stability results if we regard (2-1-1) to be the stationary equation of the corresponding diffusion problem with no flux boundary conditions, just for the very simple reason that there are no nonconstant stable stationary states for this problem (see [14]). So in this sense all bifurcating solutions are unstable. But the direction of branches does tell "how unstable" they are. This has been used by Brunowský and Fiedler

([4]) to calculate possible orbit connections between steady states. Their results together with theorems 2.1.2 and 2.1.3 imply that all orbit connections between steady states in the bifurcating branches can be given if f is sub-, superlinear, an A-B- or a C-function.

Asymptotic properties of Neumann-time-maps can be derived from paragraph 1.6, we will not mention results explicitly.

There is a strong connection between the Neumann-time-map and the period map:

If we define the period map of (2-1-3) to be the map

$$a \longmapsto \text{ the least period of } U(\cdot, a)$$

then this one is defined whenever $\mathcal{T}(a)$ exists, and the period map is just twice the Neumann-time-map because of the symmetry of solutions.

So it looks like this is all we have to say about period maps. But in order not to keep this chapter too short we will treat some more general problem.

By now you will have observed that almost everything follows so nicely because the problem $u'' + f(u) = 0$ has a first integral. So the natural generalization would be to treat first order Hamiltonian systems in the plane. Since this again is a little bit too general we restrict ourselves to problems of the form

$$(2\text{-}1\text{-}15) \qquad\qquad u' = -f_2(v) \qquad v' = f_1(u).$$

In some applications it is important to know the period map of a system like (2-1-15), for instance for bifurcation of subharmonic solutions (see [7]), or for the existence of a branch of period-four-solutions to certain time delay equations (see e.g. [18])

2.2 The period map formula

Let us consider (2-1-15) with f_i being defined on some open interval J_i and locally Lipschitz-continuous (say), in order to have existence and uniqueness. We want to find a part in the phase plane which consists of periodic orbits of (2-1-15). This is possible if we have a stationary point which is a center. So let $f_1(r_1) = f_2(r_2) = 0$. After a shift in u and v which does not affect periods we can always assume that $r_1 = r_2 = 0$. Define

$$(2\text{-}2\text{-}1) \qquad F_i(w) := \int_0^w f_i(s)\,ds \quad \text{for } w \in J_i,$$

then

$$(2\text{-}2\text{-}2) \qquad H(u,v) := F_1(u) + F_2(v) \equiv const$$

along solutions (u,v) of (2-1-15); orbits are level lines of the *Hamiltonian* H. Closed level lines of H which do not contain stationary points ,i.e. critical points of H, are periodic orbits. These occur in a neighbourhood of $(0,0)$ if $(0,0)$ is a strict local maximum or a strict local minimum of H, thus we have to assume something like positivity or negativity of the Hessian of H in $(0,0)$. Without loss of generality let us consider the case of a strict local minimum. This follows if $f_i'(0) > 0$ for $i = 1, 2$.

Let $]a_i^-, 0[$ be the maximal interval on which f_i is negative, and $]0, a_i^+[$ the maximal interval on which f_i is positive. By $\mathcal{C}$ we denote the connected component of initial values of nonconstant periodic solutions having $(0,0)$ in its closure. Then the following argument shows that $\mathcal{C}$ is contained in the rectangle

$$R := \{(u,v) \in \mathbb{R}^2 \mid a_1^- < u < a_1^+,\, a_2^- < v < a_2^+\}.$$

If $R = J_1 \times J_2$ then there is nothing to show. In the other case let

$$(2\text{-}2\text{-}3) \qquad l := \min(F_1(a_1^+), F_1(a_1^-), F_2(a_2^+), F_2(a_2^-)).$$

Then $0 < l < \infty$, and without loss of generality we can assume that

$$l = F_1(a_1^+).$$

($a_1^+ = \infty$ is possible.) Then with (2-2-2)

$$(2\text{-}2\text{-}4) \qquad H(u, a_2^{+,-}) > l \quad \text{for } u \in \,]a_1^-, a_1^+[\,\setminus\{0\}$$

and

(2-2-5) $H(a_1^{+,-}, v) > l$ for $v \in \,]a_2^-, a_2^+[\, \setminus \{0\}$

because of the definition of $]a_i^-, a_i^+[$.

Since also $\langle \nabla H(u,v), (u,v) \rangle$ is positive in $R \setminus \{(0,0)\}$ and $H(0,0) = 0 < l$ the l-level-set of H contains a simple closed curve Γ which is contained in the closure of R and touches the boundary of R at most in points $(0, a_2^{+,-})$ or $(a_1^{+,-})$ and at least in $(a_1^+, 0)$. (If $a_1^+ = \infty$ then it is "almost closed" and approaches $(\infty, 0)$ from the first and the fourth quadrant. As a still more weird case it might happen that $F_i(a_i^{+,-}) = l$ with $|a_i| = \infty$ and $a_i^{+,-} \neq a_1^+$. Then Γ will still only have one point "leaks" at infinity, and the argument below works.)

Figure 2.1.2 shows two extreme cases which can happen for Γ:

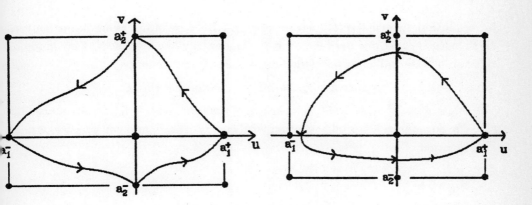

Figure 2.2.1

Any solution of (2-1-15) starting on Γ either approximates a stationary point of (2-1-15) on ∂R or a point which is not contained in the definition set of (2-1-15). Thus there are no periodic initial conditions on Γ other than stationary ones. Hence Γ is a separatrix and $\mathcal{C}$ is contained in the interior of Γ.

Actually because of (2-2-3) any level set of H inside Γ is a closed curve and either equal to $\{(0,0)\}$ or free from critical points of H. So $\mathcal{C}$ is equal to the interior of Γ minus $\{(0,0)\}$.

So it suffices to consider (f_1, f_2) on R and the natural assumptions on f_1, f_2 are the same as (1-0-1):

(2-2-6) $\qquad f_i(w) = w f_i^+(w) \quad , \quad a_i^- < 0 < a_i^+$

$\qquad\qquad$ and $\quad f_i^+ :]a_i^-, a_i^+[\to \mathbb{R}^+ \quad$ locally Lipschitz-continuous.

In 2.1 we have defined the period map as a function of a with

$$u' = -v \qquad v' = f(u)$$

$$u(0) = a \qquad v(0) = 0.$$

For periodic solutions it is more convenient to take the energy of the solution as a parameter which is the level E of the Hamiltonian along the orbit:

(2-2-7) $\qquad\qquad H(u(t), v(t)) = F_1(u(t)) + F_2(u(t)) \equiv E$

for a solution $t \mapsto (u(t), v(t))$ with energy E. Because of the assumptions on the f_i there is at most one periodic orbit for any energy level E of H thus modulo shifts in t a periodic solution is uniquely defined by its energy.

We are going to introduce a slightly different parameter though:

You might guess that we want to make use of the functions $\Phi(f_i)$ as introduced in (1-1-4). So we take a periodic solution with energy $1/2\, p^2$ and consider its period as a function of p:

Let

(2-2-8) $\quad D(\Pi) := \{p > 0 \mid$ (2-1-15) has a periodic solution with energy $\dfrac{1}{2} p^2 \}$,

and for $p \in D(\Pi)$ let $t \mapsto (U(\cdot, p), V(\cdot, p))$ be any periodic solution with energy $1/2\, p^2$ (e.g. the one with $v(0) = 0$, $u(0) > 0$, $F_1(u(0)) = 1/2\, p^2$).

Then the period map $\Pi : D(\Pi) \to \mathbb{R}^+$ is well defined by

(2-2-9) $\qquad\qquad \Pi(p) := $ the least period of $(U(\cdot, p), V(\cdot, p))$.

We have seen above that periodic orbits fill the interior of Γ, the level curve of H to the level

(2-2-10) $\qquad\qquad l = \min(F_1(a_1^+), F_1(a_1^-), F_2(a_2^+), F_2(a_2^-))$.

Thus l is the supremum of energies of periodic orbits. Hence

(2-2-11) $$D(\Pi) = \,]0, \bar{b}[\quad \text{with} \quad \bar{b} := \sqrt{2l}.$$

With transformations similar to the ones in paragraph 1.1 we can deduce a period map formula, using

(2-2-12) $$g_1 := \Phi(f_1) \qquad g_2 = \Phi(f_2)$$

as defined by (1-1-4), i.e.:

$$g_i : \,]b_i^-, b_i^+[\,\to\,]a_i^-, a_i^+[\,,$$

$$b_i^\pm = \pm\sqrt{2F_i(a_i^\pm)},$$

$$F_i(g(x)) = \frac{1}{2}x^2 \quad, \quad xg_i(x) > 0 \text{ for } x \neq 0.$$

The geometric idea behind this is to transform level lines of H into circles.

The period map formula we end up with below has been proved by Waldvogel in [35], we just use a slightly different way of deriving it.

Let (u, v) be a periodic solution of (2-1-15) with energy $1/2\,p^2$, $p > 0$. Then, modulo a shift in t, (u, v) solves the problem

(2-2-13)
$$u' = -f_2(v) \qquad v' = f_1(u)$$
$$u(0) = g_1(p) \qquad v(0) = 0.$$

With

$$u = g_1(x) \qquad v = g_2(y)$$

(2-2-13) becomes

(2-2-14)
$$g_2'(y)g_1'(x)x' = -y \qquad g_2'(y)g_1'(x)y' = x$$
$$x(0) = p \qquad y(0) = 0$$

since

$$g_1'(x)f_1(g_1(x)) = x \qquad g_2'(y)f_2(g_2(y)) = y.$$

So again there happens to be the same factor in front of both x' and y', so we can transform time to get a linear equation:

Let $t(\theta)$ be the unique solution of

$$\frac{dt}{d\theta} = g_2'(y(t))g_1'(x(t))$$

$$t(0) = 0.$$

Then, considering x, y as functions of θ, we derive from (- -)

$$\frac{dx}{d\theta} = -y \qquad \frac{dy}{d\theta} = x$$
$$x(0) = p \qquad y(0) = 0.$$

So we get

$$x(t(\theta)) = p\cos\theta \qquad y(t(\theta)) = p\sin\theta,$$

$$t(\theta) = \int_0^\theta g_1'(p\cos\varphi)g_2'(p\sin\varphi)\,d\varphi,$$

$$u(t(\theta)) = g_1(p\cos\theta) \qquad v(t(\theta)) = g_2(p\sin\theta).$$

Since g_1, g_2 are both strictly increasing the least period of (u,v) has to be $t(2\pi)$.

Thus we have the period map formula

(2-2-15) $$\Pi(p) = \int_0^{2\pi} g_1'(p\cos\theta)g_2'(p\sin\theta)\,d\theta$$

which is defined for $p \in D(\Pi) = \,]0,\bar{b}[\,$.

If we assume the right regularity for f_1, f_2 (see 1.1) then we can derive formulas for the derivatives of Π and for $\Pi^{(k)}(0)$:

(2-2-16) $$\Pi(0) = \int_0^{2\pi} g_1'(0)g_2'(0)\,d\theta = \frac{2\pi}{\sqrt{f_1'(0)f_2'(0)}}$$

(2-2-17) $\quad \Pi'(p) = \displaystyle\int_0^{2\pi} g_1''(p\cos\theta)g_2'(p\sin\theta)\cos\theta + g_1'(p\cos\theta)g_2''(p\sin\theta)\sin\theta\,d\theta$

(2-2-18) $$\Pi'(0) = \int_0^{2\pi} g_1''(0)g_2'(0)\cos\theta + g_1'(0)g_2''(0)\sin\theta\,d\theta = 0$$

$$\Pi''(p) = \int_0^{2\pi} g_1'''(p\cos\theta)g_2'(p\sin\theta)\cos^2\theta \, d\theta$$

(2-2-19)
$$+ \int_0^{2\pi} g_1'(p\cos\theta)g_2'''(p\sin\theta)\sin^2\theta \, d\theta$$

$$+ \int_0^{2\pi} 2g_1''(p\cos\theta)g_2''(p\sin\theta)\sin\theta\cos\theta \, d\theta$$

$$\Pi''(0) = g_1'''(0)g_2'(0)\int_0^{2\pi}\cos^2\theta \, d\theta + g_1'(0)g_2'''(0)\int_0^{2\pi}\sin^2\theta \, d\theta$$

(2-2-20)
$$+ 2g_1''(0)g_2''(0)\int_0^{2\pi}\sin\theta\cos\theta \, d\theta$$

$$= \pi(g_1'''(0)g_2'(0) + g_1'(0)g_2'''(0))$$

2.3 Monotonic period maps

Now we can try to get analogues of theorems 2.1.1 and 2.1.2 for the period map.

A theorem corresponding to 2.1.1 has been proved in [21]:

If f_1 and f_2 are both sub- (super-)linear then the period map is strictly increasing (decreasing).

So we concentrate on proving 2.1.2 for the period map.

First of all we can write (2-2-16) in a different way:

$$
\begin{aligned}
\Pi'(p) &= \int_0^{2\pi} g_1''(p\cos\theta)g_2'(p\sin\theta)\cos\theta + g_1'(p\cos\theta)g_2''(p\sin\theta)\sin\theta \, d\theta \\
&= \frac{1}{p}\int_0^{2\pi} g_1''(p\cos\theta)\frac{d}{d\theta}g_2(p\sin\theta)\, d\theta + \frac{1}{p}\int_0^{2\pi} g_2''(p\sin\theta)(-\frac{d}{d\theta}g_1(p\cos\theta))\, d\theta \\
&= \int_0^{2\pi} g_1'''(p\cos\theta)\sin\theta g_2(p\sin\theta)\, d\theta + \int_0^{2\pi} g_2'''(p\sin\theta)\cos\theta g_1(p\cos\theta)\, d\theta \, .
\end{aligned}
$$

Since we know that $xg_i(x) > 0$ for $x \neq 0$, the positivity of $\Pi'(p)$ ($p > 0$) follows from

(2-3-1) $\qquad g_1''' \geq 0 \qquad g_2''' \geq 0 \qquad g_i''' > 0$ for $i = 1$ or 2 ,

and Π' is negative for

(2-3-2) $\qquad g_1''' \leq 0 \qquad g_2''' \leq 0 \qquad g_i''' < 0$ for $i = 1$ or 2 .

So we get monotonicity for the period map if either both f_1 and f_2 are A-B-functions or they are both C-functions.

Actually $\Pi' > 0$ follows from assumptions which are a little more general:

2.3.1 THEOREM. *Let (2-1-2) hold for f_1 , f_2 , and let f_1 , f_2 be continuously differentiable.*

Assume that there are

$$ c_i^- \in [a_i^-, 0[\qquad c_i^+ \in]0, a_i^+] $$

with

(2-3-3) $\qquad f_i' > 0$ on $]c_i^-, c_i^+[$, $\quad f_i' \leq 0$ on $]a_i^-, c_i^-[\cap]c_i^+, a_i^+[$

for $i = 1, 2$, and further that

(2-3-4) $\qquad\qquad f_i$ is an A-function on $]c_i^-, c_i^+[$

for $i = 1, 2$.

Then

$$\Pi'(p) > 0 \quad for \quad p \in]0, \bar{b}[.$$

PROOF: With (2-2-16)

$$\Pi'(p) = \int_{-\pi/2}^{3\pi/2} g_1''(p\cos\theta)g_2'(p\sin\theta)\cos\theta\,d\theta + \int_0^{2\pi} g_1'(p\cos\theta)g_2''(p\sin\theta)\sin\theta\,d\theta$$

since it does not matter over which period we integrate.

The first integral we split into one from $-\pi/2$ to $\pi/2$ and one from $\pi/2$ to $3\pi/2$. Then with $\theta = \pi - \varphi$

$$\int_{\pi/2}^{3\pi/2} g_1''(p\cos\theta)g_2'(p\sin\theta)\cos\theta\,d\theta = -\int_{-\pi/2}^{\pi/2} g_1''(-p\cos\varphi)g_1'(p\sin\varphi)(-\cos\varphi)\,d\varphi$$

since $\cos(\pi - \varphi) = -\cos\varphi$, $\sin(\pi - \varphi) = \sin\varphi$.

The second integral we split into one from 0 to π and one from π to 2π and get with $\theta = 2\pi - \varphi$

$$\int_\pi^{2\pi} g_1(p\cos\theta)g_2''(p\sin\theta)\sin\theta\,d\theta = \int_0^\pi g_1'(p\cos\varphi)g_2''(-p\sin\varphi)(-\sin\varphi)\,d\varphi.$$

Thus

$$\Pi'(p) = \int_{-\pi/2}^{\pi/2} (g_1''(p\cos\theta) - g_1''(-p\cos\theta))g_2'(p\sin\theta)\cos\theta\,d\theta$$
$$+ \int_0^\pi (g_2''(p\sin\theta) - g_2''(-p\sin\theta))g_1'(p\cos\theta)\sin\theta\,d\theta.$$

Thus for $\Pi' > 0$ it suffices to show:

(2-3-5) $\qquad$ $g_1''(x) - g_1''(-x) > 0 \quad$ for $x \in]0, \bar{b}[$
$\qquad\qquad\quad$ $g_2''(y) - g_2''(-y) > 0 \quad$ for $y \in]0, \bar{b}[.$

So let $i \in \{1, 2\}$, $f := f_i$, $g := g_i$ etc., and define

$$]d^-, d^+[:= g^{-1}(]c^-, c^+[).$$

Since f is an A-function on $]c^-, c^+[$ we have that g''' exists on $]d^-, d^+[$, and as in the proof of theorem 1.4.2 we can show that

(2-3-6) $g''' > 0$ on $]d^-, d^+[$.

So

$$g''(x) - g''(-x) > 0 \quad \text{for } x \in]0, \bar{d}]$$

with

$$\bar{d} := \min(d^+, -d^-).$$

If $\bar{d} \geq \bar{b}$ we are finished. So let us consider $\bar{d} < \bar{b}$.

In the case that $\bar{d} < d^+$ we have $d^- = -\bar{d} > -\bar{b} \geq b^-$. So then for $x \in]\bar{d}, \min(\bar{b}, d^+)]$

$$g''(x) > g''(\bar{d}) > g''(-\bar{d}) = g''(d^-).$$

It then suffices to show that

(2-3-7) $g''(-x) \leq g''(d^-)$ for $-x < d^-$.

Let $-x < d^-$ and $u = g(-x)$. Then $u \in]a^-, c^-[$ and thus $f'|_{[u,c^-]} \leq 0$ by assumption, as well as $f|_{[u,c^-]} < 0$.

With (1-1-14) we then get

$$g''(-x) = \frac{f^2 - 2Ff'}{f^3}(u) \leq \frac{1}{f}(u) \leq \frac{1}{f}(c^-)$$

$$= \frac{f^2 - 2Ff'}{f^3}(c^-) = g''(d^-).$$

So (2-3-7) follows.

As a whole we have now shown (2-3-5) on $]0, \min(\bar{b}, d^+)[$. If $d^+ \geq \bar{b}$ we are finished again. So assume $d^+ < \bar{b} \leq b^+$, and let $x \in]d^+, \bar{b}[$.

Then $u = g(x)$ is in the interval $]c^+, a^+[$ on which $f' \leq 0$, $f > 0$. So again

$$g''(x) = \frac{f^2 - wFf'}{f^3}(u) \geq \frac{1}{f}(u) \geq \frac{1}{f}(c^+)$$

$$= \frac{f^2 - 2Ff'}{f^3}(c^+) = g''(d^+).$$

If $d^+ = \bar{d}$ then it is possible that $-x \geq d^-$. Then because of (2-3-6)

$$g''(d^+) > g''(-x).$$

For $d^+ > \bar{d}$ we get $-x < -\bar{d} = d^-$ and

$$g''(d^+) > g''(d^-) \geq g''(-x)$$

by (2-3-6) and (2-3-7).

So, all cases considered we have shown that (2-3-5) holds, and thus the theorem is proved. ∎

Remark:

The proof of theorem 2.3.1 also works in the case that $f_2(v) = v$ since then $g_2'' \equiv 0$. So for theorem 2.1.3 (i) it suffices to assume the slightly more general conditions of theorem 2.3.1 for $f_1 = f$.

2.4 Examples and applications

2.4.1 EXAMPLE. With theorem 2.3.1 we are able to recover results by Rothe ([25])
and Waldvogel ([36]) about the Lotka-Volterra-system

(2-4-1) $$x' = \alpha x(1 - y) \qquad y' = \beta y(x - 1)$$

with $\alpha, \beta > 0$, which is meant to be a simple model for a predator-prey-ecosystem,
$x \geq 0$ resp. $y \geq 0$ being the prey- resp. predator-density at time t. For some
background see e.g. [2].

With $x = e^{-u}$, $y = e^{-v}$ (2-4-1) becomes

(2-4-2) $$u' = -\alpha(1 - e^{-v}) \qquad v' - \beta(1 - e^{-u}),$$

a system of the form (2-1-15) with

$$f_1(u) = \beta(1 - e^{-u}) \qquad f_2(v) = \alpha(1 - e^{-v})$$

satisfying (2-2-6).

Morover it is trivial the f_i are A-functions on $\mathbb{R}$ where the f_i' are positive. Thus
the period map of (2-4-2) is strictly increasing.

We get the same for (2-4-1) if we take some parameter for the argument of the period map which depends monotonicly on the energy of the corresponding solution
of (2-4-2), i.e. $a > 0$ with $x(0) = a$, $y(0) = 0$ being the initial condition of the
periodic solution.

As an application one can use the results on subharmonic bifurcation in [7] in order
to decide whether some of the periodic solutions of (2-4-1) still persist when the
system is perturbed by a small nonautonomous term which is t-periodic (seasonal
influx). ∎

Instead of (2-4-2) let us consider a more general system:

(2-4-3) $$x' = -\frac{f_2(y)}{\bar{h}_1(x)} \qquad y' = \frac{f_1(x)}{\bar{h}_2(y)}$$

with a stationary point (r_1, r_2), $\bar{h}_i(r_i) > 0$.

With

$$\overline{H}_i(z) = \int_{r_i}^{z} \bar{h}_i(s) \, ds$$

we can make the transformation

$$u = \overline{H}_1(x) \qquad v = \overline{H}_2(y)$$

on the maximal intervals about r_i where $\bar{h}_i$ are positive and get

(2-4-4) $$u' = -f_2(\overline{H}_2^{-1}(v)) \qquad v' = f_1(\overline{H}_1^{-1}(u)).$$

If f_i satisfies (2-1-2) on some interval containing r_i for $i = 1, 2$ then (2-2-6) holds for

$$\tilde{f}_1 := f_1 \circ \overline{H}_1^{-1} \qquad \tilde{f}_2 := f_2 \circ \overline{H}_2^{-1}$$

on some maximal intervals $]a_i^-, a_i^+[$ containing 0.

Let us check under what assumptions on f_i and $\bar{h}_i$ the conditions of theorem 2.3.1 will hold for $\tilde{f}_i$:

(2-3-3) will be true if for f_1, f_2 we have

$$f_i'(z) = 0 \Rightarrow f_i f_i''(z) < 0.$$

For $\tilde{f}_i$ to be an A-function it suffices f or $\tilde{f}_i$ to have negative Schwarzians, i.e. with lemma 1.5.7 we need e.g.

$$S f_i \leq 0 \qquad S \overline{H}_i^{-1} < 0.$$

We have to calculate $S\overline{H}_i^{-1}$ to get conditions on $\bar{h}_i$:

Since $\overline{H}_i^{-1}(\overline{H}_i(z)) = z$

$$\bar{h}_i^2(z)(S\overline{H}_i^{-1})(\overline{H}_i(z)) + S\overline{H}_i(z) = 0.$$

So the condition for $S\overline{H}_i^{-1} < 0$ is

(2-4-5) $$S\overline{H}_i = \frac{\bar{h}_i \bar{h}_i'' - \frac{3}{2}(\bar{h}_i')^2}{\bar{h}_i^2} > 0.$$

If we want to write (2-4-3) as

(2-4-6) $$x' = -h_1(x)f_2(y) \qquad y' = h_2(y)f_1(x)$$

then with $\bar{h}_i := 1/h_i$ (2-4-5) becomes a condition on h_i, namely

$$h_i h_i'' - \frac{1}{2}(h_i')^2 < 0.$$

We collect results in the following proposition:

2.4.2 PROPOSITION. *Let* $f_i :]a_i^-, a_i^+[\to \mathbb{R}$ *be three times continuously differentiable for* $i = 1, 2$ *with*

(2-4-7) $\qquad\qquad r_i \in]a_i^-, a_i^+[\qquad f_i(r_i) = 0 \qquad f_i'(r_i) > 0,$

(2-4-8) $\qquad\qquad f_i|_{]a_i^-, 0[} < 0 \qquad f_i|_{]0, a_i^+[} > 0,$

(2-4-9) $\qquad\qquad f_i'(z) = 0 \Rightarrow f_i f_i''(z) < 0,$

(2-4-10) $\qquad\qquad S f_i \le 0$

and let $h_i :]a_i^-, a_i^+[\to \mathbb{R}^+$ *be twice continuously differentiable for* $i = 1, 2$ **with**

(2-4-11) $\qquad\qquad h_i h_i'' - \dfrac{1}{2}(h_i')^2 < 0.$

Then there exists a maximal subinterval $]0, \bar{a}[$ *of* $]a_1^-, a_1^+[$ *such that any solution of (2-4-6) with* $x(0) = a \in]0, \bar{a}[$, $y(0) = 0$ *is periodic. If* $\Pi(a)$ *is the least period of this solution then*

$$\Pi'(a) > 0 \quad \text{for } a \in]0, \bar{a}[.$$

2.4.3 EXAMPLE The following equations arise from a mathematical model describing the conflict of the sexes concerning parental investment (see [28]):

(2-4-12) $\qquad x' = -x(1-x)((a+b)y - a) \qquad y' = y(1-y)((c+d)x - c)$

where x, y are restricted to $]0, 1[$, $x(y)$ ist the proportion of females (males) persuing one of two possible strategies. a, b, c, d arise from payoff matrices. A family of periodic orbits occurs in the case a, b, c, $d > 0$.

With

$$f_1(x) = (c+d)x - c \qquad f_2(y) = (a+b)y - b \qquad h_i(z) = z(1-z)$$

(2-4-12) is of the form (2-4-6) and f_i, h_i satisfy assumptions (2-4-7) − (2-4-9) on $]a_i^-, a_i^+[=]0, 1[$ with $r_1 = c/(c+d)$, $r_2 = a/(a+b)$. Moreover $S f_i = 0$ and (2-4-11) holds for h_i since $h_i'' < 0$. So periods of (2-4-12) increase with the amplitude of solutions. ∎

2.5 Semilinear Dirichlet Problems

Instead of calculating the period map of (2-1-15) one could also give formulas for the time the system needs to pass through either of the four quadrants in the phase plane, by simply changing the bounds of the integral in (2-2-15) to 0, $\pi/2$; $\pi/2$, π etc..

A problem like this one comes up in a natural way if one considers Dirichlet problems in "divergence form":

(2-5-1)
$$(h(u'))' + f(u) = 0$$
$$u(0) = u(\lambda) = 0$$

with

(2-5-2)
$$h :]a_2^-, a_2^+[\to \mathbb{R} \quad \text{continously differentiable}$$
$$\text{with} \quad h(0) = 0, h' > 0 \quad \text{on }]a_2^-, a_2^+[$$

and

(2-5-3)
$$f :]a_1^-, a_1^+[\to \mathbb{R} \quad \text{continuously differentiable}$$
$$\text{with} \quad f|_{]a_1^-,0[} < 0, f|_{]0,a_1^+[} > 0, f'(0) > 0.$$

The equation of (2-5-1) has a unique solution $U(\cdot, p)$ for any initial condition $u(0) = 0$, $u'(0) = p \in]a_2^-, a_2^+[$. As in chapter I we can define $T(p)$ to be the first positive zero of $U(\cdot, p)$, if it exists. The graph of T then here too gives the branch of positive and negative solutions of (2-5-1) projected to the $(\lambda, u'(0))$-plane.

With $v := -h(u')$ the initial value problem of (2-5-1) can be written as

(2-5-4)
$$u' = h^{-1}(-v) \qquad v' = f(u)$$
$$u(0) = 0 \qquad v(0) = -h(p)$$

which is a problem of the form (2-1-15) with

(2-5-5)
$$f_1(u) := f(u) \qquad f_2(v) := -h^{-1}(v).$$

If $1/2\, q^2$ is the energy of the solution to (2-5-4) then we have the relation

(2-5-6)
$$\frac{1}{2}q^2 = F_2(v(0)) = F_2(-h(p)).$$

$T(p)$ is just the time the solution of (2-5-4) needs to pass through the fourth and first ($p > 0$) or through the second and third ($p < 0$) quadrant in the phase plane. Thus because of (2-5-6)

(2-5-7) $$T(p) = R(q) := \int_{-\pi/2}^{\pi/2} g_1'(q\cos\theta)g_2'(q\sin\theta)\,d\theta$$

with

(2-5-8) $$h(p) = -g_2(-q))$$

i.e. q and p depend in a strictly increasing way upon one another. The definition set of R is an interval $]q^-,q^+[$ with

(2-5-9)
$$q^- = \max(-\sqrt{2F_1(a_1^-)}, -\sqrt{2F_2(a_2^-)}, -\sqrt{F_2(a_2^+)})$$
$$q^+ = \min(\sqrt{2F_1(a_1^+)}, \sqrt{2F_2(a_2^-)}, \sqrt{F_2(a_2^+)}).$$

This is the maximal interval for which the integrand of (2-5-7) is defined. Corresponding to this is a definition interval $]b^-,b^+[$ for T.

If we could show that $R''(q)$ is of one sign then this would imply a result similar to theorem 1.4.2, since then T could have at most one critical point. But in the formula for R'' (see (2-2-19)) we get a term

$$\int_{-\pi/2}^{\pi/2} g_1''(q\cos\theta)g_2''(q\sin\theta)\cos\theta\sin\theta\,d\theta$$

which vanishes in the case $h(z) = z$ but causes trouble otherwise. The condition that g_i''' is of one sign does not suffice to get the sign of this integral e.g. via integration by parts. The obvious condition to add is that $g_1''(0) = 0$ which then implies that $xg_1''(x)$ is of one sign for $x \neq 0$. But since then $R'(0) = 0$ we get that automatically $qR'(q)$ is of one sign. So it looks like we are not able to get anything but a generalization of theorem 1.3.1.

If we have

(2-5-10) $$xg_1''(x) > (<)0 \quad \text{for } x \neq 0$$

and

(2-5-11) $$g_2'''(y) \geq (\leq)0$$

then

$$qR'(q) = \int_{-\pi/2}^{\pi/2} q\cos\theta g_1''(q\cos\theta) + g_1'(q\cos\theta)g_2''(q\sin\theta)q\sin\theta\,d\theta$$

$$> (<) \int_{-\pi/2}^{\pi/2} g_1'(q\cos\theta)g_2''(q\sin\theta)q\sin\theta\,d\theta$$

$$= \int_0^{\pi/2} q\sin\theta g'(q\cos\theta)(g_2''(q\sin\theta) - g_2''(-q\sin\theta))\,d\theta$$

$$\geq (\leq)\,0\,.$$

(2-5-11) for " $>$ " follows if f_2 is an A-B-function, i.e. an A-function since $f_2' > 0$ by the assumptions on h. " $<$ "follows from f_2 being a C-function. By (2-5-5) we can translate this into a condition on h which is

$$h'h''' - \frac{4}{3}(h'')^2 > (<)0\,.$$

$xg_1''(x) > (<)0$ follows from conditions on $f = f_1$ as the ones in theorem 1.3.1. So here is what we have proved:

2.5.1 THEOREM. *Let* $]b^-, b^+[\ni p \mapsto T(p)$ *be the first Dirichlet-time-map of*

$$(h(u'))' + f(u) = 0$$
$$u(0) = 0 \quad u'(0) = p$$

with (2-5-2) *and* (2-5-3) *holding for* h *and* f, *and let* f *be twice* , h *three times continuously differentiable.*

If

(2-5-12) $$u f''(u) < 0 \quad \text{for all } u \neq 0 \text{ with } f'(u) > 0$$

and if

(2-5-13) $$h'h''' - \frac{4}{3}(h'')^2 > 0$$

then

$$pT'(p) > 0 \quad \text{for all } p \in]b^-, b^+[\ \backslash\ \{0\}\,.$$

If

(2-5-14) $$u f''(u) > 0 \quad \text{for } u \neq 0$$

and if

(2-5-15) $$h'h''' - \frac{4}{3}(h'')^2 < 0$$

then

$$pT'(p) < 0 \quad \text{for all } p \in \,]b^-, b^+[\,\setminus\, \{0\}\,.$$

Bifurcation points- and directions can be calculated as in paragraph 1.2. The methods of calculating the asymptotic behaviour of T can also be adopted, though there are so many cases to consider in this more general setting that we omit details.

CHAPTER III

GENERALIZATIONS

3.1 Positive solution branches of Dirichlet problems with $f \geq 0$

In chapter I we have restricted f so that $f(0) = 0$, $f'(0) > 0$. If f has a multiple zero at 0 then it is well known that no finite-λ-bifurcation occurs but branches "bifurcate" from $(+\infty, 0)$. We can handle this case too. On the other hand f' can have a singularity at 0, but still the Dirichlet problem might make sense. As a third more general case f can be different from 0 at 0. Then we do not have the trivial solution branch $(\lambda, 0)$, but still branches of nontrivial solutions occur which can intersect in points (λ, u_0) with $u'_0(0) = u'_0(\lambda) = 0$.

We will treat all these more general cases, but will only consider branches of positive solutions. With a change of variables results can be applied to negative solutions as well, and by adding up time maps one can derive some results for branches of sign changing solutions.

Let us postpone the case $f(0) < 0$ for a while and assume that f is positive on some interval $]0, a[$. We will now try to make the "most general" assumptions on f so that the time map is still defined. For this we use some Sobolev spaces whose definitions can looked up in the standard literature.

First of all the initial value problem

$$
\begin{aligned}
u'' + f(u) &= 0 \\
u(0) &= 0 \; u'(0) = p
\end{aligned}
$$

(3-1-1)

has to have a unique solution for $p > 0$.

We want to be able to integrate (3-1-1), thus we have to assume that f is locally integrable in $[0, a[$.

Then if u is a solution of (3-1-1) in $H^{2,1}_{loc}([0, t_0[)$ with $u' > 0$ on $[0, t_0[$ then u solves

$$
\begin{aligned}
u' &= \sqrt{p^2 - 2F(u)} \\
u(0) &= 0
\end{aligned}
$$

(3-1-2)

on $[0, t_0[$. On the other hand by Caratheodory's theorem ([10]) there exists a solution of (3-1-2) whose interval of definition is the maximal interval $[0, t_0[$ with

$F(u(t)) < 1/2 \, p^2$ for all $t \in [0, t_0[$. Since the right hand side of (3-1-2) is strictly decreasing in u this solution is unique and is a $H_{loc}^{2,1}([0, t_0[)$-solution of (3-1-1).

For the Dirichlet time map of (3-1-1) to be defined in $p > 0$ it is necessary that for the solution of (3-1-2) there is a finite time $t_0 > 0$ with $u'(t_0) = 0$, i.e. $F(u(t_0)) = 1/2 \, p^2$. So we have to assume that $p \in]0, \sqrt{2F(a)}[$. t_0 can be calculated by

$$(3\text{-}1\text{-}3) \qquad t_0 = \int_0^{t_0} 1 \, dt = \int_0^{g(p)} \frac{du}{\sqrt{p^2 - 2F(u)}}$$

via $u = u(t)$ with $g = \Phi(f)$ defined as usual by $F(g(x)) = 1/2 \, x^2$.

The integral in (3-1-3) becomes finite if we assume that f is locally bounded away from 0 in $]0, a[$, i.e. $1/f \in L_{loc}^\infty(]0, a[)$:

Let $1/f \le c$ in $[\epsilon, g(p)]$ with $0 < \epsilon < g(p)$, $0 < c < \infty$. Then

$$\int_0^{g(p)} \frac{du}{\sqrt{p^2 - 2F(u)}} = \int_0^{\epsilon} \frac{du}{\sqrt{p^2 - 2F(u)}} + \int_{\epsilon}^{g(p)} \frac{du}{\sqrt{p^2 - 2F(u)}}$$

$$\le \frac{\epsilon}{\sqrt{p^2 - 2F(\epsilon)}} + c \int_0^{g(p)} \frac{f(u)}{\sqrt{p^2 - 2F(u)}} \, du$$

$$= \frac{\epsilon}{\sqrt{p^2 - 2F(\epsilon)}} + cp < \infty.$$

Having a finite time t_0 with $u'(t_0) = 0$, $u(t_0) = g(p)$, we can extend u as a solution of (3-1-1) by inflection in t_0. The extension is unique since f is bounded away from 0 near $g(p)$, thus any solution of (3-1-1) for $t > t_0$ has to solve

$$u' = -\sqrt{p^2 - 2F(u)}$$

$$u(t_0) = g(p)$$

which again admits a unique solution.

The found solution can easily be shown to be in $H^{2,1}([0, 2t_0])$ and its first positive zero is

$$T(p) = 2t_0 = 2 \int_0^{g(p)} \frac{du}{\sqrt{p^2 - 2F(u)}} = 2 \int_0^p \frac{g'(x)}{\sqrt{p^2 - x^2}} \, dx$$

via the transformation $u = g(x)$. The transformation is allowed since g' is integrable near 0 and bounded near p. With $x = p \sin \theta$ we get the usual time map formula.

So we take f from the following classes of nonlinearities $\mathcal{F}_a$, $a > 0$:

$$(3\text{-}1\text{-}4) \qquad \mathcal{F}_a := \{ f \in L_{loc}^1([0, a[) \mid f > 0 \text{ a.e. in }]0, a[, \frac{1}{f} \in L_{loc}^\infty(]0, a[) \},$$

and we have proved the following:

3.1.1 PROPOSITION. *For any $a > 0$ and any $f \in \mathcal{F}_a$ problem (3-1-1) has a unique $H_{loc}^{2,1}$-solution $u(t) = U(t,p)$ for any $p > 0$.*

If $b := \sqrt{2F(a)} \in {]}0, \infty]$ then

(3-1-5) $$T(p) := \min\{t > 0 \mid U(t,p) = 0\}$$

is defined for $p \in {]}0, b[$ and

(3-1-6) $$T(p) = 2 \int_0^p \frac{g'(x)}{\sqrt{p^2 - x^2}} \, dx = 2 \int_0^{\pi/2} g'(p \sin \theta) \, d\theta,$$

where $g = \Phi(f)$ is defined on $[0, b[$ by

(3-1-7) $$F(g(x)) = \frac{1}{2}x^2 \quad, \quad F(u) = \int_0^u f(s) \, ds.$$

g is contained in

$$\mathcal{G}_b := \{g \in H_{loc}^{1,1}([0, b[) \mid g(0) = 0, \, g' > 0 \text{ a.e. in } [0, b[, \, g' \in L_{loc}^\infty({]}0, b[)\}.$$

The properties of g stated in the proposition follow from the usual formula

(3-1-8) $$g'(x) = \frac{x}{f(g(x))} = \frac{\sqrt{2F(u)}}{f(u)}, \, u = g(x).$$

Despite the weak assumptions on f one can show that T is always continuous, even Hölder-continuous:

3.1.2 PROPOSITION. *Let f be in $\mathcal{F}_a$ for some $a > 0$. The the Dirichlet time map of f as defined in proposition 3.1.1 is locally Hölder-continuous with exponent $1/2$ in ${]}0, b[$.*

PROOF: Let p and q be in $[2\epsilon, b-\epsilon]$ with $\epsilon > 0$, $a < p$. Then $g'(x)/x \leq c < \infty$ in $[\epsilon, b - \epsilon]$. So by (3-1-6)

$$|T(p) - T(q)| \leq \left| \int_0^q g'(x) \left(\frac{1}{\sqrt{p^2 - x^2}} - \frac{1}{\sqrt{q^2 - x^2}} \right) + \int_q^p g'(x) \frac{1}{\sqrt{p^2 - x^2}} \right|$$

$$\leq \int_0^\epsilon g'(x) \left(\frac{1}{\sqrt{q^2 - x^2}} - \frac{1}{\sqrt{p^2 - x^2}} \right)$$

$$+ \int_\epsilon^q g'(x) \left(\frac{1}{\sqrt{q^2 - x^2}} - \frac{1}{\sqrt{p^2 - x^2}} \right) + \int_q^p g'(x) \frac{1}{\sqrt{p^2 - x^2}}$$

$$:= I_1 + I_2 + I_3.$$

The factor of $g'(x)$ in I_1 is monotonicly increasing in x thus

$$I_1 \le \int_0^\epsilon g'(x) \left(\frac{1}{\sqrt{q^2 - \epsilon^2}} - \frac{1}{\sqrt{p^2 - \epsilon^2}} \right) = g(\epsilon) \left(\frac{1}{\sqrt{q^2 - \epsilon^2}} - \frac{1}{\sqrt{p^2 - \epsilon^2}} \right)$$

$$\le dg(\epsilon)|p - q|$$

with some constant $d > 0$ since p and q are bounded away from ϵ.

Since $g'(x)/x$ is bounded in $[\epsilon, p]$ we get

$$I_2 \le c \int_\epsilon^p \frac{x}{\sqrt{q^2 - x^2}} - \frac{x}{\sqrt{p^2 - x^2}}\, dx = c \left(\sqrt{p^2 - q^2} + \sqrt{q^2 - \epsilon^2} - \sqrt{p^2 - \epsilon^2} \right)$$

$$\le d|p - q|^{1/2}$$

with some constant $d > 0$ depending on ϵ.

For I_3

$$I_3 \le c \int_q^p \frac{x}{\sqrt{p^2 - x^2}}$$

$$= c\sqrt{p^2 - q^2} \le d|p - q|^{1/2}.$$

So this shows the $C_{loc}^{0+1/2}$-regularity of T. ∎

Here as in chapter 1 the behaviour of T near $p = 0$ detemines the local bifurcation behaviour from 0 of (1-0-5) in the case $f(0) = 0$:

3.1.3 PROPOSITION. *Let f be in $\mathcal{F}_a$ for some $a > 0$.*

(i) If $f \ge c > 0$ a.e. in a neighbourhood of 0 then

$$T(p) \to 0 \quad as \quad p \to 0.$$

(ii) If f is continuous near 0 with $f(0) = 0$ and if

(3-1-9) $$\lim_{u \to 0} \frac{2F(u)}{f^2(u)} := c \in [0, \infty]$$

exists then

$$T(p) \to \pi\sqrt{c} \quad for \quad p \to 0.$$

(iii) If f is continuous near 0 with $f(0) = 0$ and there is some $\alpha > 0$ such that

(3-1-10) $$\lim_{u \to 0} \frac{u^\alpha}{f(u)} := c \in \,]0, \infty[$$

exists then

$$\lim_{p \to 0} p^{\frac{\alpha-1}{\alpha+1}} T(p) = d \in \,]0, \infty[$$

exists and can be calculated from c and α by

$$(3\text{-}1\text{-}11) \qquad d = 2 \left(\frac{2}{\alpha+1} \right)^{\frac{\alpha}{\alpha+1}} c^{\frac{1}{\alpha+1}} \int_0^{\pi/2} (\sin \theta)^{\frac{1-\alpha}{1+\alpha}} \, d\theta.$$

PROOF: (i) and (ii) immediately follow from (3-1-6), (3-1-8) and Lebesgue's theorem.

For (iii) we first show that $\lim_{x \to 0} x^\beta g'(x)$ exists for $\beta = (\alpha-1)/(\alpha+1)$: With $u = g(x)$

$$x^\beta g'(x) = \frac{(2F(u))^{(\beta+1)/2}}{f(u)} = \frac{(2F(u))^{\alpha/(\alpha+1)}}{f(u)} = \frac{u^\alpha}{f(u)} \left(\frac{2F(u)}{u^{\alpha+1}} \right)^{\alpha/(\alpha+1)}.$$

By de l'Hôpital's theorem

$$\lim_{u \to 0} \frac{2F(u)}{u^{\alpha+1}} = \frac{2}{\alpha+1} \lim_{u \to 0} \frac{f(u)}{u^\alpha} = \frac{2}{\alpha+1} c^{-1}.$$

Thus

$$\lim_{x \to 0} x^\beta g'(x) = \left(\frac{2}{\alpha+1} \right)^{\frac{\alpha}{\alpha+1}} c^{\frac{2}{\alpha+1}}.$$

With

$$p^\beta T(p) = 2 \int_0^{\pi/2} (p \sin \theta)^\beta g'(p \sin \theta)(\sin \theta)^{-\beta} \, d\theta$$

the assertion follows from Lebesgue's theorem (note that $\beta < 1$). ∎

Some remarks:

The assumption of (ii) follows of course if the assumption of (iii) is satisfied with $\alpha = 1$. If f has a multiple zero at 0 then de l'Hôpital's theorem applied to (3-1-9) gives that $T(p) \to \infty$ as $p \to 0$, the branch bifurcates from $(\lambda = \infty, u \equiv 0)$. (iii) then gives the order of growth of T near $p = 0$. Note that T will always be integrable, the "area under the branch" is always finite.

If $f'(u) \to \infty$ as $u \to 0$ then by (iii) $T(0) = 0$, the branch bifurcates from $(0,0)$.

One interesting phenomenon can be derived from this proposition: If $f(u) = uf^+(u)$ with f^+ being positive on $]a^-, a^+[$ with $a^- < 0 < a^+$ then the time maps of f can be calculated by proposition 3.1.1 by considering $f|_{]0,a^+[}$ and $f|_{]a^-,0[}$ with a suitable transformation separately. If f^+ has a jump at $u = 0$ then the first Dirichlet time map has to have a jump, too, using (ii). Instead of having one bifurcation point for positive and negative solutions as in Chapter I this one splits into two separate ones. But any even numbered branch is continuous in 0 since it is given by $T_{2i}(p) = i(T(p) + T(-p))$. Odd numbered branches given by $T_{2i+1}(p) = (i + 1)T(p) + iT(-p)$ again have a discontinuity at $p = 0$. For similar local bifurcation phenomena see [29].

For the more general assumptions of this paragraph the asymptotic behaviour of time maps is the same as stated in proposition 1.6.1. For the sake of better reference we will state this proposition again, also in some of the proofs Lebuege's theorem cannot be so readily applied since now g' is no longer bounded near 0.

3.1.4 PROPOSITION. *Let f be in $\mathcal{F}_a$ for some $a > 0$.*

(i) Assume that

(3-1-12)
$$a < \infty \quad , \quad \lim_{u \to a} f(u) = 0$$

and

(3-1-13)
$$f(u) \leq c(a - u)(\ln(a - u))^2$$

for u near a with some constant $c > 0$. Then

$$T(p) \to \infty \text{ as } p \to b.$$

(ii) Assume that

(3-1-14)
$$a < \infty \quad , \quad \lim_{u \to a} f(u) = 0$$

and

(3-1-15)
$$f(u) \geq c(a - u)(-\ln(a - u))^q$$

for u near a with constants $c > 0$, $q > 2$.

Then $T(p)$ has a positive and finite limit as $p \to b$.

(iii) Let

(3-1-16)
$$a < \infty \quad , \quad \lim_{u \to a} f(u) = \infty$$

and

(3-1-17) $F(a) = \infty$, f increasing near a .

Then
$$T(p) \to 0 \text{ as } p \to b = \infty .$$

(iv) *Assume that*

(3-1-18) $a = \infty$, $\lim_{u \to a} \dfrac{2F(u)}{f^2(u)} = c \in [0, \infty]$.

Then
$$T(p) \to \pi \sqrt{c} \text{ for } p \to b .$$

PROOF: (i) and (ii) follow exactly as in the proof of proposition 1.6.1.

Ad (iii): First we note that the case $F(a) < \infty$ is already contained in the more general assumptions on f, for then we could extend f to a function in $\mathcal{F}_{a+\epsilon}$ with some $\epsilon > 0$.

If $F(a) = \infty$ then as in 1.6.1 we can show that
$$\lim_{x \to \infty} g'(x) = 0 .$$

Thus
$$\int_{\pi/4}^{\pi/2} g'(p \sin \theta) \, d\theta$$

becomes small for p large. For the rest
$$\int_0^{\pi/4} g'(p \sin \theta) \, d\theta \leq \sqrt{2} \int_0^{\pi/4} g'(p \sin \theta) \cos \theta \, d\theta = \sqrt{2} \frac{g(p \sin(\pi/4))}{p} \to 0$$

as $p \to \infty$.

Ad (iv): If $c = \infty$ then g' and thus T approaches ∞ as $p \to \infty$. For $c = \infty$ $g'(x)$ stays bounded for large x, hence by Lebegue's theorem
$$\lim_{p \to \infty} \int_{\pi/4}^{\pi/2} g'(p \sin \theta) \, d\theta = \frac{\pi}{4} \sqrt{c} .$$

For the rest
$$\int_0^{\pi/4} g'(p \sin \theta) \, d\theta = \frac{1}{p} \int_0^{\pi/2} \frac{d}{d\theta} g(p \sin \theta)(\cos \theta)^{-1}$$
$$= \frac{g(p/\sqrt{2})}{p/\sqrt{2}} - \frac{1}{p} \int_0^{\pi/4} g(p \sin \theta) \frac{d}{d\theta} (\cos \theta)^{-1} \, d\theta$$
$$=: I_1 - I_2 .$$

Since $g'(x) \to \sqrt{c}$ as $x \to \infty$ we get

$$I_1 \to \sqrt{c} \text{ for } p \to \infty.$$

Since $g(p\sin\theta)/p \leq g(p)/p$ is bounded indepently of p large we can apply Lebegue's theorem to I_2, and since $\lim_{p\to\infty} g'(p\sin\theta)/p = \sqrt{c}\sin\theta$

$$I_2 \to \sqrt{c} \int_0^{\pi/4} \sin\theta \frac{d}{d\theta}(\cos\theta)^{-1}\, d\theta = \sqrt{c} - \sqrt{c}\frac{\pi}{4}.$$

With this (iv) and the proposition is proved. ∎

With formula (3-1-6) we can again differentiate T, provided the derivative of the integrand with respect to p exists and is integrable. We collect some results in the next proposition:

3.1.5 PROPOSITION. *Let $f \in \mathcal{F}_a$ be n-times weakly differentiable in $]0, a[$ with locally bounded derivatives,i.e. $g^{(n+1)}$ exists in the weak sense on $]0, b[$ and is locally bounded.*

(i) *If the function $x \mapsto x^n g^{(n+1)}(x)$ is in $L^1_{loc}([0, b[)$ then $T^{(n)}$ exists on $]0, b[$ with*

(3-1-19) $$T^{(n)}(p) = 2 \int_0^{\pi/2} (\sin\theta)^n g^{(n+1)}(p\sin\theta)\, d\theta.$$

(ii) *Assume that $g^{(n+1)}$ does not change sign in a neighbourhood of 0. Then*

$$x^k g^{(k)}(x) \to 0 \text{ as } x \to 0 \text{ for } k = 0, 1, ..., n$$

and

$$x \mapsto x^n g^{(n+1)}(x) \text{ is in } L^1_{loc}([0, b[).$$

(iii) *For $n = 1, 2$ we ge the well known formulas*

(3-1-20) $$g''(x) = \frac{\sqrt{f^2 - 2Ff'}}{f^3}(u), \ u = g(x),$$

(3-1-21) $$g'''(x) = g'(x)\frac{h(u)}{f^4(u)}, \ u = g(x)$$

with

(3-1-22) $$h = -3f'(f^2 - 2Ff') - 2Fff'' .$$

(iv) In $n = 1$ *,* f'' *exists in a neighbourhood of* 0 *and does not change sign then* g'' *does not change sign in a neighbourhood of* 0 *, i.e.* T' *exists on* $]0, b[$ *.*

(v) If $n = 2$ *,* f''' *exists in a neighbourhood of* 0 *and if the functions* f' *and* $f'f''' - 5/3(f'')^2$ *do not change sign near* 0 *then* g''' *has no change of sign near* 0 *, i.e.* T'' *exists on* $]0, b[$ *.*

(Zeroes do not count as sign changes.)

PROOF: (i) is clear.

Ad (ii): We prove the assertion by induction:

$n = 0$ is clear.

$n - 1 \to n$:

If $g^{(n+1)}$ is of one sign near 0 then the same holds for $g^{(k)}$ for all $k \leq n$. Thus $\lim_{x \to 0} x^k f^{(k)}(x) = 0$ for all $k \leq n-1$ and $n^{n-1} g^{(n)}(x)$ is integrable near 0 by the induction hypothesis. Now

$$x^n g^{(n+1)}(x) = \frac{d}{dx}(x^n g^{(n)}(x) + r(x))$$

with $r(x)$ being the sum of terms $const \, x^k g^{(k)}(x)$, $k \leq n-1$, i.e. $r(x) \to 0$ as $x \to 0$.

Since $g^{(n+1)}$ is of one sign near 0 the limit for $x \to 0$ of $x^n g^{(n)}(x) + r(x)$, i.e. of $x^n g^{(n)}(x)$ has to exist. The limit can be nothing but 0 for otherwise $x \mapsto x^{n-1} g^{(n)}(x)$ could not be integrable.

Thus $x^n g^{(n+1)}(x) = d/dx \, R(x)$ with $\lim_{x \to 0} R(x) = 0$. By the theorem of monotone convergence we get herefrom that $x^n g^{(n+1)}(x)$ is L^1 near 0.

(iii) is clear by implicit differentiation, which is also allowed in the weak sense.

Ad (iv) We have to show that g'' has a single sign near 0. This is true since the sign of g'' is the same as of $f^2 - 2Ff'$ which is a monotonic function by assumption.

Ad (v) The sign of g''' is the same as the one of h (see (3-1-21), (3-1-22)). As in the proof of theorem 1.4.2

$$3f'h' = 5f''h'' + R(u)$$

with

$$R = 6Ff\left(\frac{5}{3}(f'')^2 - f'f'''\right)$$

having one sign near 0 by assumption. Since

$$\frac{d}{du}\frac{h^3}{(f')^5} = \frac{h^2}{(f')^6}R$$

the function $h^3/(f')^5$ is monotonic and h does not change sign in a neighbourhood of 0. ∎

That $\lim_{x\to 0} x^n g^{(n)}(x) = 0$ is of course necessary for $x^n g^{(n+1)}$ to be integrable near 0.

On the other hand one could think of examples of g where g'' exists, yet T' cannot exist because $xg''(x)$ is not integrable. This is the case if g' is oscillating too strongly (i.e. the leading term of g' is $\sin(1/x) + const$). For this it is again necessary that f' oscillates strongly near 0.

One could of course give more general criteria as in (iv), (v) for xg'', $x^2 g'''$ to be in $L^1_{loc}([0, b[)$, e.g. limiting the amount of oscillation of f', f'', but this does not seem the right place for writing down lengthy conditions.

In analogy to the result of proposition 3.1.2 the $C^{n+1/2}_{loc}$-regularity of T can be proved if f satisfies the assumptions of proposition 3.1.5.

Now we are ready to prove generalizations of theorems 1.3.1 and 1.4.2:

3.1.6 THEOREM. Let $f \in \mathcal{F}_a$ for some $a > 0$ and let f be continuously differentiable on $]0, a]$.

(i) Assume that f'' exists on intervals where f' is positive with

(3-1-23) $f''(u) < 0$ for all u with $f'(u) > 0$.

Then

$$T'(p) > 0 \text{ on }]0, b[.$$

(ii) Assume that $\lim_{u\to 0} f(u) = 0$ and that f'' exists on $]0, a[$ with

(3-1-24) $f''(u) > 0$ for all $u \in]0, a[$.

Then

$$T'(p) < 0 \text{ on }]0, b[.$$

PROOF: Ad (i): First of all it follows from the assumptions that f' cannot change sign from $-$ to $+$ on $]0, a[$. If now $f' \leq 0$ near 0 then it stays so for the rest of the interval and the result follows trivially since g'' and $f^2 - 2Ff'$ have the same sign. (This case of course is only possible for $f(0) > 0$.)

Now let f' be positive near 0. Then f' must have a limit in $]0, \infty]$ and f a limit in $[0, \infty[$ as u approaches 0. Thus f' is decreasing near 0 and an L^1-function. Herefrom it follows that $\lim_{u \to 0} u f'(u) = 0$, and thus

$$F(u)f'(u) = \frac{F(u)}{u} u f'(u) \to 0 \text{ as } u \to 0$$

since $F(u)/u$ is bounded together with f. Hence

$$\lim_{u \to 0}(f^2 - 2Ff')(u) \geq 0$$

and

$$\frac{d}{du}(f^2 - 2Ff')(u) = -2Ff''(u) > 0$$

as long as $f'(u) > 0$. Thus $f^2 - 2Ff'$ is positive for $f' > 0$, and trivially for $f' \leq 0$. So g'' and thus T' is positive.

Ad (ii): From the assumptions it follows that $f' > 0$ on $]0, a[$ and that f' has a finite limit as $u \to 0$. Thus $(f^2 - 2Ff')(u) \to 0$ as $u \to 0$ and its derivative is negative. So g'' and thus T' is always negative. ∎

Note that we cannot say anything in the case $f(0) > 0$, $f'' > 0$, since we then only know that g'' starts off positive and has at most one change of sign from $+$ to $-$. So T' starts positive (of course, since $T(0) = 0$) and might become negative later, but may have more than one change of sign, or less (see the beginning of 3.3).

The generalization of theorem 1.4.2 follows from the following lemmata:

3.1.7. LEMMA. *Let* $f \in \mathcal{F}_a$ *be an A-B-function on* $]0, a[$ *(see definition 1.4.1). Then* $\lim_{u \to 0} f(u)$ *and* $\lim_{u \to 0} f'(u)$ *exist, and we have the following:*

 (i) *g''' has at most one change of sign in $]0, b[$ which then is from $-$ to $+$.*

 (ii) *If $f(0) > 0$ and $f'(0) > 0$ then g''' is negative near 0.*

 (iii) *If $f(0) > 0$ and $f'(0) \leq 0$ then g''' is positive on $]0, b[$.*

(iv) If $f(0) = 0$ then g''' is positive on $]0, b[$.

PROOF: If f is an A-B-function then there is a $c \in [0, a]$ sucht that $f' > 0$ on $]0, c[$, $f' < 0$ on $]c, a[$, one of the intervals possibly being empty. So f will always be continuous in 0. On $]0, c[$ by assumption

$$(3\text{-}1\text{-}25) \qquad \frac{d^2}{du^2}(f')^{-2/3} = \frac{2}{3}(f')^{-8/3}\left(\frac{5}{3}(f'')^2 - f'f'''\right) > 0$$

and on $]c, a[$

$$(3\text{-}1\text{-}26) \qquad \frac{d^2}{du^2}f^{-2} = 2f^{-4}(3(f')^2 - ff'') > 0.$$

So in any case f' will be continuous in 0.

Ad (i): g''' has the same sign as h defined by (3-1-22). As in the proof of theorem 1.4.2

$$3f'h' > 5f''h \text{ on }]0, c[.$$

Thus

$$(3\text{-}1\text{-}27) \qquad \frac{d}{du}h(f')^{-5/3} = \frac{1}{3}(f')^{-8/3}(3f'h' - 5f''h) > 0 \text{ on }]0, c[.$$

So h can have at most one change of sign in $]0, c[$ which then is from $-$ to $+$. On $]c, a[$

$$(3\text{-}1\text{-}28) \qquad h = -3f'f^2 + 2F(3(f')^2 - ff'') > 0$$

by assumption.

Ad (ii): If $f(0) > 0$, $f'(0) > 0$ then $]0, c[\neq \emptyset$ and $f(0)$ must be finite. Also $f'(0) < \infty$ for otherwise $(f')^{-2/3}(0) = 0$ with $(f')^{-2/3}(u) \leq du$ with a constant $0 < d < \infty$ by (3-1-25). Then it would follow that $f'(u) \geq du^{-3/2}$ with some constant $d > 0$, a contradiction to f' being integrable.

As a result

$$(3\text{-}1\text{-}29) \qquad 0 < f(0) < \infty \quad , \quad 0 < f'(0) < \infty.$$

From (3-1-25), (3-1-29) it follows that $\lim_{u \to 0} uf''(u) = 0$, and herefrom again that $\lim_{u \to 0} F(u)f''(u) = 0$. Hence

$$h(u) = (-3f'f^2 + 6F(f')^2 - 2Fff'')(u) \to -3(f'f^2)(0) < 0 \text{ as } u \to 0$$

and g''' is negative near 0.

(iii) is trivial by (3-1-28).

Ad (iv): If $f(0) = 0$ then f' must be positive near 0, and we get $f'(0) < \infty$ as in (ii).

As a first case we consider

$$0 < f'(0) < \infty.$$

Then as in (ii)

$$\lim_{u \to 0} h(u) = 0 = \lim_{u \to 0} (h(f')^{-5/3})(u).$$

Because of (3-1-27) then h is positive near 0.

In the case

$$f'(0) = 0$$

we get

$$(g'(x))^2 = \frac{2F(u)}{f^2(u)} \to \infty \text{ as } u = g(x) \to 0.$$

So $\liminf_{x \to 0} g''(x) = -\infty$, $\limsup_{x \to 0} g'''(x) = +\infty$. Since g''' can only change sign once we get $g''' > 0$ near 0 and on all of $]0, b[$. ∎

3.1.8 LEMMA. *If* $f \in \mathcal{F}_a$ *is a C-function on* $]0, a[$ *(see definition 1.4.1) then* $\lim_{u \to 0} f(u)$ *and* $\lim_{u \to 0} f'(u)$ *exist and* f' *is either positive or negative on* $]0, a[$.

 (i) *If* $f' > 0$ *then* $g''' < 0$ *on* $]0, b[$.

 (ii) *If* $f' < 0$ *then* $g''' > 0$ *on* $]0, b[$.

PROOF: If f is a C-function on $]0, a[$ then

(3-1-30) $$\frac{d^2}{du^2} |f'|^{-2/3} = \frac{2}{3} |f'|^{-8/3} \left(\frac{5}{3} (f'')^2 - f'f''' \right) < 0$$

on intervals where $|f'| > 0$.

This way f' cannot have any zeroes on $]0, a[$ and also $f'(0)$ must be nonzero.

Ad (i): If $f' > 0$ then as in lemma 3.1.7 we get

(3-1-31) $$\frac{d}{du} h(f')^{-5/3} < 0.$$

With the same sort of arguments as in the proof of lemma 3.1.7 (ii) $F(u)f'(u) \to 0$ and $F(u)f''(u)(f')^{-3/5} \to 0$ as $u \to 0$. Thus

$$h(f')^{-5/3} = -3f^2(f')^{-2/3} + 6F(f')^{1/3} - 2Fff''(f')^{-5/3}$$

has a nonpositive limit as $u \to 0$. Hence $g''' < 0$ on $]0, b[$.

Ad (ii): If $f' < 0$ then

$$\frac{d}{du} h(f')^{-5/3} > 0.$$

In the case $f(0) < \infty$ we get as before that $\lim_{u \to 0}(h(-f')^{-5/3})(u) \geq 0$, thus $g''' > 0$ on $]0, b[$.

If $f(0) = \infty$ then $g'(x) \to 0$ and $g'(x)/x \to 0$ as $x \to 0$. g''' has a sign near 0, thus $\lim_{x \to 0} g''(x)$ esists which then must be 0. Since g' is positive g''' must be positive near 0 and on $]0, b[$. ∎

The following theorem now follows easily:

3.1.9 THEOREM. *Let f be a function in $\mathcal{F}_a$, $a > 0$.*

(i) If f is an A-B-function on $]0, a[$ with $f(0) = 0$ or $f(0) > 0$ and $f'(0) < 0$ then

$$T''(p) > 0 \text{ on }]0, b[.$$

(ii) If f is a C-function on $]0, a[$ then

$$T''(p) < 0 \text{ on }]0, b[\text{ if } f' > 0 \text{ on }]0, a[$$

and

$$T''(p) > 0 \text{ on }]0, b[\text{ if } f' < 0 \text{ on }]0, a[.$$

Examples are discussed in the next paragraph.

3.2 Applications and Examples

3.2.1 EXAMPLE. In the first example we can see what happens to the bifurcating positive solution branch of some polynomial f without complex zeroes during a homotopy which starts with the usual situation that $f(0) = 0$, $f'(0) > 0$ and then shifts the function to the left so that $f(0)$ becomes positive. The starting picture we know from chapter 1. If in this situation $f''(0) \leq 0$ then after a small shift $f''(0)$ must be negative, and we are in the situation where theorem 3.1.6 can be applied, telling that $T'(p) > 0$ in this case with $T(0) = 0$, the branch "bifurcates" from $(0,0)$. (There is of course no real bifurcation since by now the trivial solution branch $(\lambda \in \mathbb{R}^+, u \equiv 0)$ has disappeared.) As long as $f'(0) > 0$ we cannot tell anything about the sign of T'' besides from that it starts off negative. After a bigger shift $f'(0)$ becomes negative and then we know that T'' will always be positive, which of course implies that T' is always positive.

If in the starting situation $f''(0) > 0$ then we can tell less, but it is the more interesting case. As a reminder here is the picture of the positive solution branch in this case:

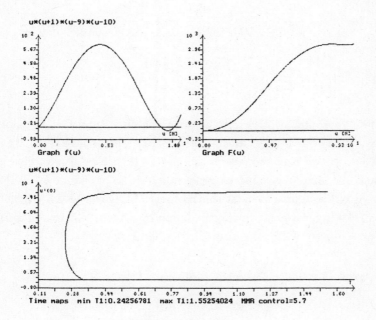

Figure 3.2.1

Now f is shifted a little bit to the left. Then the solution set must be close to the one for $f(0) = 0$. Hence, since $T(0) = 0$ necessarily T' must have at least two changes of sign for the shift being small enough (see figure 3.2.2).

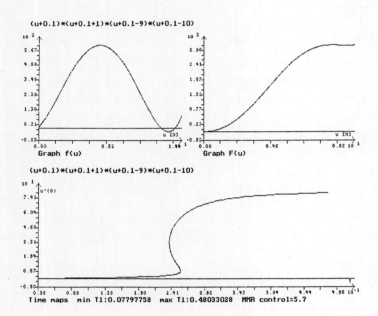

Figure 3.2.2

Smoller and Wasserman ([31]) have studied the time map for f being a cubic polynomial with two real roots and with $f''' < 0$. They could prove that T' also can have at most two sign changes in the case $f(0) > 0$. All numerical examples suggest that one should be able to generalize this to f being an A-B-function, but a proof is not available to us at this moment. In figures (3.2.3)-(3.3.4) you can further persue what happens during the homotopy, these last two pictures can be proved since in these $f''(0) < 0$ resp. $f'(0) < 0$. ∎

3.2.2 EXAMPLE. Let us consider the problem

(3-2-1)
$$v_t = v_{xx} - (v_x)^2 + v^q \quad , v > 0$$
$$v(t,0) = v(t,\lambda) = 0.$$

It has been studied for existence of solutions which blow up in finite time ([6]).

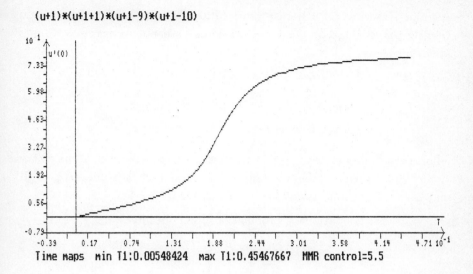

Figure 3.2.3

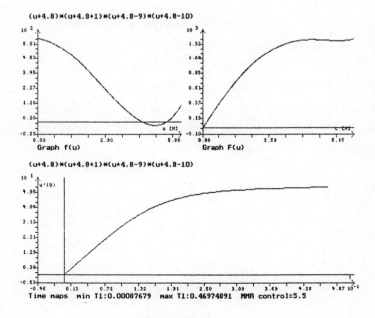

Figure 3.2.4

Peletier and Kawohl ([20]) discuss the relation of this phenomenon to another one, namely the existence of dead core stationary states (see example 1.6.2) of an equivalent problem:

With $w := e^{-v}$ (3-2-1) becomes equivalent to

(3-2-2)
$$w_t = w_{xx} - w(-\ln w)^q \quad , 0 < w < 1$$
$$w(t,0) = w(t,\lambda) = 1 .$$

A dead core stationary state of (3-2-2) corresponds to a "solution" of the stationary equation of (3-2-1) which takes on infinite values on a whole subinterval of $]0, \lambda[$ and may give rise to finite time blow up solutions of (3-2-1).

With $u := 1 - w$ the stationary problem of (3-2-2) becomes

(3-2-3)
$$u'' + f(u) = 0$$
$$u(0) = u(\lambda) = 0$$

with

(3-2-4)
$$f(u) := (1 - u)(-\ln(1 - u))^q$$

satisfying the assumptions of the last paragraph. As in example 1.6.2 the existence of dead core solutions of (3-2-3) is equivalent to $T(p)$ approaching a finite value as $p \to b := \sqrt{2F(1)}$, T being the Dirichlet time map of f.

Proposition 3.1.4 now tells us that this is the case iff $q > 2$.

Moreover, using theorems 3.1.6, 3.1.9 we can say (at least for some q) something about the number and the stability of stationary states of (3-2-1), (3-2-2):

Using $x := (-(\ln(1 - u))$ we get

(3-2-5)
$$f(u) = e^{-x} x^q$$

(3-2-6)
$$f'(u) = x^{q-1}(q - x)$$

(3-2-7)
$$f''(u) = e^x x^{q-2} q(q - 1 - x)$$

(3-2-8)
$$f'''(u) = e^{2x} x^{q-3} q((q - 1)(q - 2) - x^2) .$$

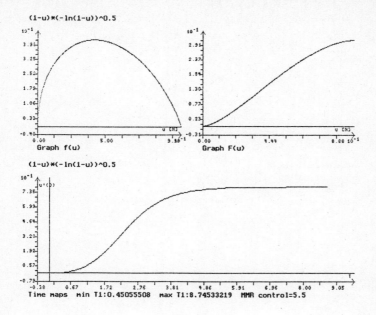

Figure 3.2.5

For the local bifurcation we thus get a finite positive bifurcation point for $q = 1$ ($f'(0) = 1$), bifurcation from infinity for $q > 1$ ($f'(0) = 0$)) and bifurcation from $(0,0)$ for $q < 1$ ($f'(0) = \infty$).

Since $f'' < 0$ for $q < 1$ we can use theorem 3.1.6 to derive that $T'(p)$ is always positive in this case. As a result (3-2-1), (3-2-2) has a unique stationary state for each λ which is stable. (It is clear that stable solutions of (3-2-3) correspond to stable stationary states of (3-2-1), (3-2-2) and vice versa.)

If now $1 \le q \le 2$ then f''' is negative and it is trivial that $f'f''' - 5/3(f'')^2$ is negative on the $f' > 0$-interval. For the B-function property we calculate

$$(ff'' - 3(f')^2)(u) = -x^{2q-2}(3x^2 - (6 - q)x + q(2q + 1)),$$

and the term in brackets is positive for $q \ge 1$.

Hence f is an A-B-function on $]0,1[$ for $1 \le q \le 2$ from which with theorem 3.1.9 it follows that $T''(p) > 0$. Together with the asymptotic behaviour of T we get the result that in this case there is a value $\lambda^* > 0$ such that (3-2-1), (3-2-2) has no nontrivial stationary solution for $\lambda < \lambda^*$ and for $\lambda > \lambda^*$ there is exactly

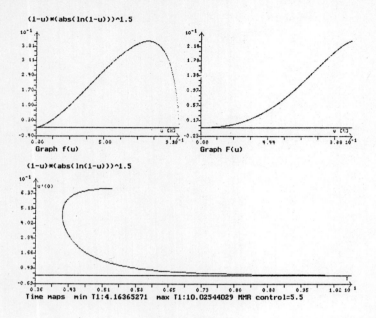

Figure 3.2.6

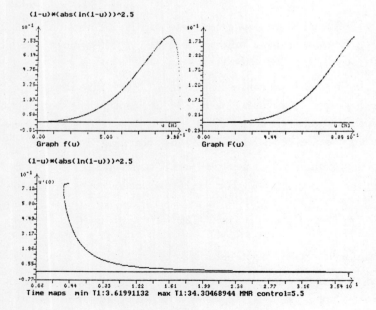

Figure 3.2.7

one small (big for (3-2-2)) unstable solution and one big (small for (3-2-2)) stable solution.

One can also show that f is an A-B-function for e.g. $q \leq 2.3$, but certainly not for $q \geq 2.4$, but this result is not worth the effort it takes typing the proof. Anyhow, the numerical pictures in figures 3.2.5–3.2.7 suggest that $T'' > 0$ holds for all $q > 1$.

Note that the bifurcation picture of (3-2-3) gives exactly the stationary solutions of (3-2-1) projected to the $(\lambda, v'(0))$-plane, since $v = -\ln(1 - u)$. ∎

We can further apply the results of the last paragraph if we consider reaction diffusion equations with a reaction term $r(v)$, v being the density of the substance, and a diffusion coefficient $d = d(v)$ which depends on v, i.e. the diffusivity of the substance changes with its density. As a stationary equation for space dimension one we then have

$$(d(v)v')' + r(v) = 0$$
$$v(0) = v(\lambda) = 0$$

(3-2-9)

with some positive function $d(v)$ which is –say– C^1 and integrable.

With

$$D(v) := \int_0^v d(s)\, ds$$

(3-2-10)

this problem becomes equivalent to one for $u := D(v)$:

$$u'' + r(D^{-1}(u)) = 0$$
$$u(0) = u(\lambda) = 0\,.$$

(3-2-11)

Functions $d(v) = v^m$, $m > -1$ often occur in models. If $m > 0$ this models a substance whose particles have little movement if there are very few of them and whose diffusion velocity increases with the number of particles present in the space element, modelling e.g. flows through porous media. The case $m < 0$ models a situation where particles have a very high velocity if there are few around and and a very low one if their density is large, thus modelling some effect of stickiness.

In this case the situation that the substance is produced by a constant rate, $r(v) = \alpha v$, $\alpha > 0$ is very easy since we are able to calculate the time map explicitly:

Let c denote any positive constant which only depends on m, α. Then from $u = D(v) = cv^{m+1}$ we get $f(u) = cu^{1/(m+1)}$, $F(u) = cu^{(2+m)/(1+m)}$ and thus via $F(g(x)) = 1/2\,x^2$ that $g'(x) = cx^{m/(m+2)}$. So by (3-1-6)

$$(3\text{-}2\text{-}12) \qquad\qquad T(p) = cp^{\frac{m}{m+2}}\,.$$

As expected the pictures for $m > 0$ and $m < 0$ look completely different. For $m > 0$ a stable branch of nontrivial solutions bifurcates from $(0,0)$. The zero-solution has to be unstable since it can only be stable before the first bifurcation point; this makes sense since diffusion for small u-values is not enough to to stabilize 0. If $m < 0$ then we have an unstable nonzero solution for all $\lambda > 0$. 0 is always stable. The unstable solution should play the role of a "hairtrigger" which decides whether the time dependent system is smoothed out to 0 by fast diffusion or else goes to ∞ because of increasing stickiness.

In what we have just said we were very sloppy, since we have argumented as if the time dependent system would be governed by the equation

$$(3\text{-}2\text{-}13) \qquad\qquad u_t = u_{xx} + r(D^{-1}(u))$$

though one really has to consider

$$(3\text{-}2\text{-}14) \qquad\qquad (D^{-1}(u))_t = u_{xx} + r(D^{-1}(u))\,.$$

But since one can use the same Liapounov functional for both equations one can easily see by variational methods that linearized stability of (3-2-13) implies asymptotic stability for (3-2-14).

With $f = r \circ D^{-1}$ it is easy to write down conditions on r and D such that f satisfies one of the assumptions of theorems 3.1.6, 3.1.9 using the results of section 1.5. Also see the second example in the introduction. A related example will also be discussed in section 4 of this chapter.

3.3 More about the case f > 0

In the last section we have seen that in the case $f(0) > 0$ there can be more turns of the positive branch than if we take a function with similar properties and $f(0) = 0$. From lemma 3.1.7 we get that g''' does have at most one sign change if f is an A-B-function. This implies that g'' can have at most two changes of sign. If we now could prove that (see (3-1-19))

$$(3\text{-}3\text{-}1) \qquad pT'(p) = 2 \int_0^p (p^2 - x^2)^{-1/2} x g''(x)\, dx$$

has at most as many sign changes as g'' then we would be able to prove the hypothesis stated at the end of example 3.2.1 — and many more very nice ones —.

There is a whole theory about integral operators which do have this property first developed by I.J.Schoenberg. One can read all about it in the book [19] by S.Karlin. A linear integral operator with a kernel function $K(x,y)$ defined on $I \times J$

$$(3\text{-}3\text{-}2) \qquad (Lf)(x) := \int_J K(x,y) f(y)\, dy$$

is called *variation-diminishing* if Lf has at most as many sign changes on I as f has on J. For a proper definition of sign change see [19], for us the intuitive notion suffices that "touch down" zeroes do not count as sign changes.

With a change of variables the the integral operator for $pT'(p)$ becomes a convolution operator with kernel

$$(3\text{-}3\text{-}3) \qquad K(x,y) := \begin{cases} (x - y)^q & \text{for } 0 \le y < x \\ 0 & \text{for } y \ge x \end{cases}$$

defined on $\mathbb{R}^+ \times \mathbb{R}^+$ with $q = -1/2$. We then get that $\sqrt{x} T'(\sqrt{x}) = (Lf)(x)$ with

$$f(y) = g''(\sqrt{y}).$$

But the conditions given in [19] for the variation diminishing property do not apply to K if $q = -1/2$, see [19, p.109]. In fact it is not variation diminishing as a counterexample shows:

Let us first see why there has to be a counterexample. If L is given by (3-3-2) with K as in (3-3-3) with $q = -1/2$ then it is easy to see that

$$(3\text{-}3\text{-}4) \qquad L \circ Lf = cF \quad , \quad F(y) = \int_0^y f(s)\, ds,$$

c being some positive constant. If now the variation diminishing property of L was true then it would necessarily follow that Lf must always have at least as many sign changes as F. Now using integration by parts (assuming f to be regular enough such that everything exists nicely)

$$Lf(x) = \int_0^x (x-y)^{-1/2} f(y)\, dy = \int_0^x (x-y)^{-1/2} \frac{d}{dy}(F(y) - F(x))\, dy$$

(3-3-5)

$$= \frac{F(x)}{\sqrt{x}} + \int_0^x \frac{1}{2}(x-y)^{-3/2}(F(x) - F(y))\, dy\,.$$

Now let us assume that F looks like as shown in figure 3.3.1:

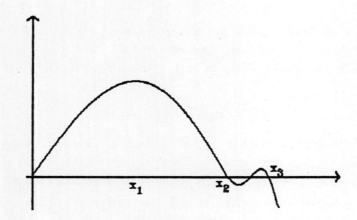

Figure 3.3.1

Then by (3-3-5) Lf will be positive up to x_1, will have become negative by x_2 and will again be negative a little bit right of x_3. If we now make the hump before x_2 big and the other ones very small then Lf will only change sign ones between 0 and x_2 and stay negative afterwards, a counterexample.

These calculations give the idea how to construct the real counterexample which is

$$f(y) := \sqrt{y}(y - 0.78)((y-2)^2 + 0.01)\,.$$

Lf can be calculated explicitly since f is the sum of powers $y^{n+1/2}$. Figure 3.3.2 shows the graph of f and the graph of Lf cut off at ± 0.5. f only changes sign once whereas Lf has three changes of sign.

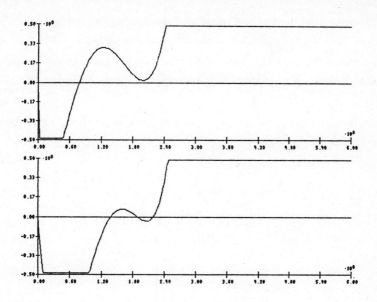

Figure 3.3.2

But it can be proved ([19, p.107]) that L is variation diminishing if the kernel K in (3-3-3) has an exponent $q > (n-1)/2$ and only f are considered which have no more than n changes of sign. So let us try to write T' with an integral operator having a kernel with a bigger q:

$$\frac{1}{2}p^3 T'(p) = \int_0^p \frac{p^2}{\sqrt{p^2-x^2}} x g''(x)\, dx$$

$$= \int_0^p \sqrt{p^2-x^2}\, x g''(x)\, dx + \int_0^p \frac{x}{\sqrt{p^2-x^2}} x^2 g''(x)\, dx.$$

Via integration by parts, provided g'' is regular enough, the second term becomes

$$\int_0^p \sqrt{p^2-x^2}(2xg''(x)+x^2 g'''(x))\, dx,$$

hence we get

(3-3-6) $$p^3 T'(p) = 2\int_0^p (p^2-x^2)^{1/2}(3xg''(x)+x^2 g'''(x))\, dx.$$

This corresponds to an exponent $q = 1/2$ in K. We could achieve exponents $q = n+1/2$ for any integer $n > 0$ if g has enough regularity. But the argument in the integral operator then becomes a more and more complicated sum of $x^j g^{(j+1)}(x)$, and at least the author could not quite see what to do with these.

Anyway, with (3-3-6) we can get hold of T' which have at most one change of sign (in the theorem the symbol f is restored to its proper meaning again):

3.3.1 THEOREM. *Let f be in $\mathcal{F}_a$, assume that g is three times continuously differentiable on $]0, b[$ where $g := \Phi(f)$.*

Further assume that the function

(3-3-7) $$\hat{g}(x) := 3xg''(x) + x^2 g'''(x)$$

changes sign no more than once on $]0, b[$.

Then for the time map T of f it follows that T' has at most as many sign changes as $\hat{g}$.

We just remark that if $\hat{g}$ has a sign near 0 then it must be a L^1-function and formula (3-3-1) for T' holds.

We will now give some condition on f such that $\hat{g}$ does have at most one sign change. It is not a very satisfactory one, but at least it gives the well known result for the Gelfand-problem (see below).

3.3.2 THEOREM. *Assume that $f \in \mathcal{F}_a$ is three times continuously differentiable on $]0, a[$ with*

(3-3-8) $$f''' \geq 0 \quad , \quad (f')^2 - ff'' \geq 0$$

on $]0, a[$.

Then T' has at most one change of sign in $]0, b[$.

PROOF: We calculate a term which dominates the sign of $\hat{g}$ ((3-3-8)):

$$x\hat{g}(x) = \frac{d}{dx}(x^3 g''(x)) = \frac{d}{dx}\left((g')^3(x)(f^2 - 2Ff')(u)\right) , \, u = g(x)$$

via (3-1-8), (3-1-20). Hence

$$x\hat{g}(x) = 3(g')^2(x)g''(x)(f^2 - 2Ff')(u) - (g')^4(x)(2Ff'')(u)$$
$$= \frac{(g')^2(x)}{f^3(u)}\left(3(f^2 - 2Ff')^2(u) - 4F^2 ff''(u)\right) .$$

$\hat{g}$ thus has the same number of sign changes as

$$h := 3(f^2 - 2Ff')^2 - 4F^2 ff''.$$

We have assumed that $f''' \geq 0$, thus there is some $c \in [0, a]$ such that

$$f''|_{]0, c[} \leq 0 \qquad f''|_{]c, a[} > 0.$$

On $]0, c[$ we have $h \geq 0$. Thus it suffices to show that $h' < 0$ on $]c, a[$.
By calculation

$$h' = -2F \left(5f''(f^2 - Ff') + 2Fff''' \right) \leq -10Ff''(f^2 - Ff')$$

since $f''' \geq 0$. Thus it suffices to show that h_1 is positive on $]c, a[$ for

$$h_1 := f^2 - Ff'.$$

From the assumptions it follows that f'/f is decreasing on $]0, a[$. So, if we assume that f' has some zero $d \in]0, a[$ then $f' \geq 0$ on $]0, d[$ and $f' \leq 0$ on $]d, a[$. This can only be true if $d \in]0, c]$. Thus $f' \leq 0$ on $]c, a[$ which implies that h_1 is positive there.

In the other case f' is either negative on $]0, a[$ and $h_1 > 0$ follows trivially or f' is positive on $]0, a[$ and

$$\frac{d}{du} \frac{h_1}{f'}(u) = \frac{f}{(f')^2}(u) \left((f')^2 - ff'' \right) (u) > 0$$

by assumption.

Since then $h_1(u)/f'(u)$ certainly has a nonnegative limit as $u \to 0$ we get that $h_1 > 0$ on $]0, a[$ in this case.

Thus $h' < 0$ on $]c, a[$ in all cases possible and the theorem is proved. ∎

As a first example let us take some polynomial f of degree ≥ 2 which has no complex zeroes and is $+\infty$ at $u = +\infty$. Also let us assume that all zeroes of f are nonpositve. If then $f(0) = 0$ with $f'(0) \geq 0$ we know what happens from the results in section 1, for then $T(0) > 0$ and $T'' > 0$, $T' < 0$ ($f'' > 0$) and $T \to 0$ as $p \to \infty$.

If we now shift f to the left such that $f(0)$ becomes positive then with the same argumentation as in example 3.2.1 there has to be at least one sign change of T'.

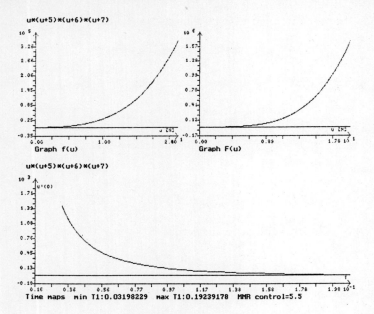

Figure 3.3.3

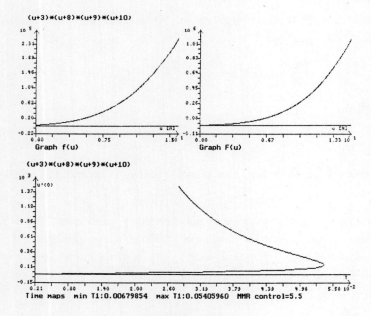

Figure 3.3.4

Since now f satisfies the assumptions of theorem 3.3.2 ($(\ln f)'' < 0$) we can also show that there is at most one change of sign for T' and the pictures of figures 3.3.3, 3.3.4 can be proved.

Note that the shifting of f to the left is equivalent to taking $f(0) = 0$ and instead of homogenous Dirichlet conditions considering $u(0) = u(\lambda) = d > 0$. One should be able to apply some of the results we got here to problems like this with d and λ being parameters (see [23]).

As a next example we get the well known result for the positive solution branch of the famous Gelfand-problem for space dimension one if we use theorem 3.3.2. In this case $f(u) = e^u$ and satisfies assumptions (3-3-8) easily. As a result the branch looks like the one in figure 3.3.4.

Note that the results in these examples could neither be derived from the fact that T'' has a single sign nor from some more general property as that $d^2/dq^2\, T(p(q))$ has a sign for some change of variables $p = p(q)$; for this expression has to be negative somewhere near the maximum of T and must become positive later since $T \to 0$ as $p \to \infty$.

3.4 Non bifurcating branches and f(0) < 0

We are now going to see what happens to bifurcating positive solution branches if some f with $f(0) = 0$, $f'(0) > 0$ is shifted to the right so that f becomes negative on an interval $]0, r[$. This way $f(u(t))$ and thus $u''(t)$ changes sign for positive solutions. It turns out that more generally we might as well consider branches of solutions for which f has at least one change of sign on the range of u. The maxima of such solutions always have to lie in an interval on which f is positive. So we look for positive solutions of

$$\begin{aligned} u'' + f(u) &= 0 \\ u(0) = u(\lambda) &= 0 \end{aligned}$$

(3-4-1)

which have maxima in some interval $]r, r + a[$ with

(3-4-2) $$r, a > 0 \quad , \quad f(r) = 0 \quad , \quad f|_{]r, r + a[} > 0 \,.$$

As usual let

(3-4-3) $$F(u) := \int_0^u f(s)\, ds$$

which now may take on negative values. Then for a solution u with maximum $q \in]r, r + a[$ we have

(3-4-4) $$\frac{1}{2}(u')^2 + F(u) \equiv F(q)$$

so it necessary that $F(u(t)) \leq F(q)$ for all $t \in [0, \lambda]$, i.e.

(3-4-5) $$\int_u^q f(s)\, ds \geq 0 \text{ for all } u \in]0, q[\,.$$

A necessary condition such that (3-4-5) holds for some maximal q-interval $]q_0, r + a[$ is (see figure 3.4.1)

(3-4-6) $$\int_u^{r+a} f(s)\, ds = F(r + a) - F(u) > 0 \text{ for all } u \in [0, r + a] \,.$$

So let us henceforth assume (3-4-2) and (3-4-6) for f.

If we now want to calculate the Dirichlet time map for solutions with maximum in $]q_0, r + a[$ then we have to start u with an initial value $u'(0) = p$ such that

(3-4-7) $$\sqrt{2F(q_0)} < p < \sqrt{2F(r + a)}$$

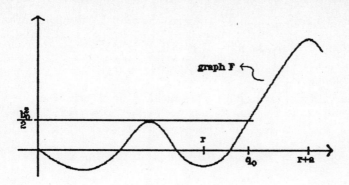

Figure 3.4.1

because of (3-4-4). Then we can argument as in 3.1:

As long as $u' > 0$ u will solve

(3-4-8)
$$u' = \sqrt{p^2 - 2F(u)}$$
$$u(0) = 0.$$

Existence and uniqueness for (3-4-8) follows if we assume that f is locally bounded in $[0, r+a[$ (note that now the right hand side of (3-4-8) is no longer monotonic in u, so we have to make some stronger assumptions). Let us for simplicity just assume that f is continuous on $[0, r+a[$.

u now will solve (3-4-8) with $u' > 0$ until $u'(t_0) = 0$, i.e. $2F(u(t_0)) = p^2$, and t_0 is again half of $T(p)$. So we get the same time map formula as in 3.1 with the difference that we can no longer define the function g.

But in order to make some use of foregoing results let us write the time map in a different way:

t_0 is the sum of two times t_1 and t_2 where t_1 is the time u needs to reach the level r, t_2 is just the time it needs afterwards to reach its maximum. t_1 can be calculated via

(3-4-9)
$$t_1 = \int_0^{t_1} 1\, dt = \int_0^r \frac{du}{\sqrt{p^2 - 2F(u)}}.$$

t_2 (see figure 3.4.2) is nothing but the time the solution of the initial value problem

(3-4-10)
$$\hat{u}'' + \hat{f}(\hat{u}) = 0$$
$$\hat{u}(0) = 0 \quad \hat{u}'(0) = u'(t_1)$$

needs to reach its maximum with $\hat{f}$ resulting from a shift of f

$$\hat{f}(\hat{u}) := f(r + \hat{u}).$$

$\hat{f}$ now is a function in $\mathcal{F}_a$ with $\hat{f}(0) = 0$, thus the time map $\hat{T}$ of $\hat{f}$ exists and

$$t_2 = \frac{1}{2}\hat{T}(u'(t_1)).$$

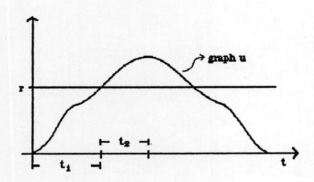

Figure 3.4.2

We can now calculate $u'(t_1)$ as a function of p via (3-4-8), thus

(3-4-11)
$$t_2 = \frac{1}{2}\hat{T}(\hat{p}) \quad , \quad \hat{p} = \sqrt{p^2 - 2F(r)}$$

So here is our result:

3.4.1 PROPOSITION. *Let f be continuous on $[0, a+r[$ with (3-4-2) and (3-4-6) holding for a, r, f. Let $]q_0, r + a[$ be the maximal subinterval of $]r, r + a[$ such that (3-4-5) holds for all $q \in]q_0, r + a[$.*

Then the Dirichlet time map of T of f is defined for all $p \in \,]b_0, b[$ with

(3-4-12) $$b_0 := \sqrt{2F(q_0)} \quad , \quad b := \sqrt{2F(a+r)},$$

and the following formula holds for T :

(3-4-13) $$T(p) = S(p) + \widehat{T}(\hat{p})$$

with

(3-4-14) $$S(p) := 2 \int_0^r \frac{du}{\sqrt{p^2 - 2F(u)}} \,,$$

(3-4-15) $$\hat{p} := \sqrt{p^2 - 2F(r)}$$

and $\widehat{T}$ being the time map of $\hat{f} \in \mathcal{F}_a$ with

(3-4-16) $$\hat{f}(\hat{u}) := f(r + \hat{u}) \,.$$

With this formula one now can get hold of $T^{(n)}(p)$ provided $\widehat{T}^{(n)}$ exists, for S is always C^∞. This again follows from $p^2 - 2F(u)$ being bounded from below by $p^2 - 2F(q_0) = p^2 - b_0^2 > 0$ on $[0, r]$. So we get

(3-4-17) $$T'(p) = S'(p) + \widehat{T}'(\hat{p}) \frac{d\hat{p}}{dp}$$

with

(3-4-18) $$S'(p) = -2 \int_0^r p(p^2 - 2F(u))^{-3/2} \, du$$

and a corresponding formula for T'' with

(3-4-19) $$S''(p) = 4 \int_0^r (p^2 + F(u))(p^2 - 2F(u))^{-5/2} \, du \,.$$

In the case that $b_0 = 0$ and $T(b_0)$ is finite the corresponding solution branch approaches a solution which has zero derivatives at both boundary points of $[0, \lambda]$. In the neighbourhood of this solution there can be four different kinds of solutions to (3-4-1): One being positive, one which has a big positive hump followed by a little negative one, one which is the inflection of the last one and one with a big positive hump in the middle and two little negative ones at both sides of the interval. So we expect to have a bifurcation point there. So let us give a condition for $T(0)$ being finite and calculate the direction of the positive solution branch:

3.4.2 PROPOSITION. *Let f satisfy the assumptions of proposition 3.4.1 with $b_0 = 0$, i.e $F(q_0) = F(0) = 0$, and let $q_0 > r$. Then necessarily $f(0) \le 0$.*

(i) If $f(0) < 0$ then $T(p)$ has a positive and finite limit as $p \to 0$.

(ii) If $f(0) = 0$ and

$$(3\text{-}4\text{-}20) \qquad f(u) \le -cu(-\ln u)^q \text{ with } c > 0, q > 2$$

then $T(p)$ has a positive and finite limit as $p \to 0$.

(iii) If $f(0) = 0$ and

$$(3\text{-}4\text{-}21) \qquad f(0) \ge -cu(-\ln u)^2 \text{ with } c > 0$$

then $T(p) \to \infty$ as $p \to 0$.

(iv) Assume that $\widehat{T}$ is differentiable and that $f(u)$ is negative for $0 < u$ near 0. Then

$$\lim_{p \to 0} T'(p) = \begin{cases} \dfrac{2}{f(0)} \text{ for } f(0) < 0 \\ -\infty \text{ for } f(0) = 0. \end{cases}$$

PROOF: From the assumptions it follows that F is negative on $]0, q_0[$, especially $F(r)$ has to be negative. But then $f(0) = F'(0)$ cannot be positive.

Moreover if $p \to 0$ then $\hat{p} \to \sqrt{-2F(r)} > 0$, thus $\widehat{T}(\hat{p})$ has a positive and finite limit as $p \to 0$. The behaviour of T at 0 thus is governed by S.

Ad (i): If $f(0) < 0$ then the function $u \mapsto (-2F(u))^{-1/2}$ is integrable on $[0, r]$, thus $S(p)$ has a positive and finite limit as $p \to 0$.

The proofs for (ii) and (iii) are analogous to the ones for theorem 3.1.6 (i), (ii).

Ad (iv): Let $f|_{]0, \epsilon]}$ be negative. We then split S' (see (3-4-18)) into the sum of S_1 and S_2 with

$$S_1(p) = -2 \int_0^\epsilon p(p^2 - 2F(u))^{-3/2} \, du$$

and

$$S_2(p) = -2 \int_\epsilon^r p(p^2 - 2F(u))^{-3/2} \, du.$$

Then $S_2(p) \to 0$ as $p \to 0$, and for S_1 we make a change of variables using the function h with

$$F(h(x)) := -\frac{1}{2}x^2 .$$

With $u = h(py)$ and $\delta := \sqrt{-2F(\epsilon)}$ we then have

$$S_1(p) = -2 \int_0^{\delta/p} \frac{h'(py)}{p}(1+y^2)^{-3/2} \, dy .$$

Now with $u = h(py)$ we get

$$\frac{h'(py)}{p} = -\frac{y}{f(u)} \to -\frac{y}{f(0)}$$

as $p \to 0$ if $f(0) < 0$. Thus in this case the limit of $S_1(p)$ for $p \to 0$ exists and is

$$\frac{2}{f(0)} \int_0^\infty y(1+y^2)^{-3/2} \, dy = \frac{2}{f(0)} .$$

For $f(0) = 0$ we can show in the same way that $S_1(p) \to -\infty$ as $p \to 0$.

For the remainder of T' we just observe that

$$\frac{d}{dp}\widehat{T}(\hat{p}) = \widehat{T}'(\hat{p})\frac{p}{\sqrt{p^2 - 2F(r)}} \to 0$$

as $p \to 0$. ∎

If $b_0 > 0$ then u will approach some homoclinic orbit as $p \to b_0$ thus T will then go to ∞, unless we have some singular situation at the homoclinic point similar to the one stated in 3.4.2 (ii). We will got go into details about this.

The behaviour of T near b is governed by the one of $\widehat{T}$ at $\hat{b} = \sqrt{b^2 - 2F(r)} = \sqrt{2\widehat{F}(a)}$ which has been discussed at length in 1.6, 3.1. So we will not dwell upon this either.

Let us now discuss what local bifurcation picture will evolve in the situation of proposition 3.4.2 if we also consider negative and sign changing solutions, i.e. if we assume f to be defined on some interval $] - a^-, r + a[$ with $a^- > 0$, $f|_{] - a^-, 0]} < 0$ and f satisfying the assumptions of 3.4.2 on $[0, r + a]$. Let us assume that f is continuously differentiable so everything we write down exists easily.

Now the time map $T(p)$ for negative p is nothing but $\overline{T}(-p)$ if $\overline{T}$ is the time map of

$$(3\text{-}4\text{-}22) \qquad\qquad \bar{f}(\bar{u}) := -f(-\bar{u}),$$

a function which is in $\mathcal{F}_{a^-}$. From the results of 3.1 we get that $\overline{T}(0) = 0$, so we have a negative solution branch starting at $(0,0)$.

As a result of proposition 3.4.2 the positive branch starts off at (λ_0, u_0) where $0 < \lambda_0 < \infty$ and u_0 is a solution with $u_0'(0) = u_0'(\lambda) = 0$. It turns to the left since $T'(0) = 2/f(0) < 0$.

Solutions with one change of sign are given by $\lambda = T(p) + T(-p)$ which also approaches λ_0 as $p \to 0$. So this branch also goes through (λ_0, u_0).

Then there is the branch which contains solutions with two sign changes and negative derivative at 0. This is given by $\lambda = 2T(p) + T(-p)$, $p < 0$ which is the same as $2\overline{T}(-p) + T(-p) \to \lambda_0$ as $p \to 0$. So it meets the same point (λ_0, u_0). Moreover its derivative at $p = 0$ is $-2\overline{T}'(0) - T'(0) = -4/\bar{f}(0) - 2/f(0) = 2/f(0)$ as can be derived from proposition 3.4.2 and the formula for $\overline{T}'$ given in 3.1. So this branch connects smoothly to the positive branch.

At (λ_0, u_0) we thus have the situation that two smooth solution curves intersect, one being made up of solutions which are symmetric in their maximum (the positive ones and the ones with two sign changes), the other one consisting of nonsymmetric solutions with a single change of sign.

Note the resemblance to the symmetry breaking bifurcation situation for $\Delta u + \lambda f(u)$ treated in [32].

The picture at (λ_0, u_0) is constantly repeated afterwards as we consider solutions with more and more sign changes.

Figure 3.4.3 shows the bifurcation picture for some polynomial f with $f(0) < 0$. The fact that the positive solution branch has but a single turn can be proved using theorem 3.4.4.

It is now very easy to give a condition for T' to be negative since this already holds for S' and $d/dp\,\widehat{T}(\hat{p})$ and $\widehat{T}'(\hat{p})$ have the same sign. So the following theorem is a consequence of theorem 3.1.6:

3.4.3 THEOREM. *Let the assumptions of proposition 3.4.1 hold for f and further assume that f is twice continuously differentiable on $]r, r+a[$ with f'' being positive there.*

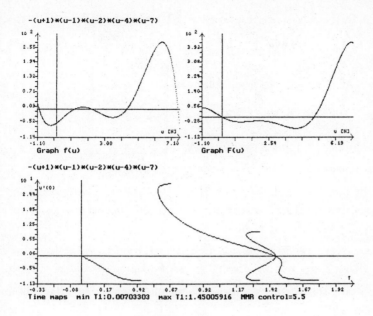

Figure 3.4.3

Then T' *is negative on* $]b_0, b[$.

The conditions are certainly satisfied if r is the largest zero of some polynomial f with no complex zeroes. In the case $b_0 > 0$ we get $T(b_0) = +\infty$, $T(b = +\infty) = 0$ and T' is negative in between. For each λ we have a maximal solution which is unstable.

Next let us try to find conditions under which a branch can only change direction once. We know conditions under which $\widehat{T}''$ is positive, so we check under what conditions we can prove that $d^2/d\hat{p}^2\, S$ is nonnegative:

Since $p^2 = \hat{p}^2 + 2F(r)$ we get $S(p) = \widehat{S}(\hat{p})$ with

$$\widehat{S}(\hat{p}) = 2 \int_0^r \left(\hat{p}^2 - 2(F(u) - F(r))\right)^{-1/2} du$$

and

$$\widehat{S}''(\hat{p}) = 4 \int_0^r \left(\hat{p}^2 + F(u) - F(r)\right) \left(\hat{p}^2 - 2(F(u) - F(r))\right)^{-5/2} du.$$

If we want to find an easy condition for $\widehat{S}''$ to be positive we have to assume that

$$\hat{p}^2 + F(u) - F(r) = p^2 + F(u) - 3F(r) > 0.$$

Since $p^2 > b_0^2 = 2F(q_0)$ this will follow if

(3-4-23) $$2(F(q_0) - F(r)) + (F(u) - F(r)) \geq 0.$$

(3-4-23) will always hold if $F(r)$ is the minimum of F on $[0, q_0]$, thus always in the bifurcation case that f is negative on $[0, r[$.

But just in the case that "f is on the borderline to having complex zeroes", i.e. if f is positive on $]0, r[$ exept for a finite number of touch down zeroes then (3-4-23) will certainly fail to hold.

(3-4-23) is a very crude condition and it might be possible to show that $\widehat{S}''$ is positive without any further conditions of F, but we could not do so.

So let us state the theorem which is a result of these calculations and theorem 3.1.9:

3.4.4 THEOREM. *Let the assumptions of proposition 3.4.1 hold for f and assume that f is an A-B-function on $]r, r + a[$.*

Further assume that (3-4-23) holds for all $u \in [0, r]$.

Then

$$\frac{d^2}{d\hat{p}^2} T(p) > 0 \quad \text{with } p^2 = \hat{p}^2 + 2F(r).$$

3.4.5 EXAMPLE. We will apply above results to a problem which has been studied more extensively by Crandall and Peletier in [12] and will recover some of their results:

Consider the stationary equation of a reaction diffusion process with density dependent diffusion coefficient $d(v) = v^m$, $m > -1$ (see section 3.2):

(3-4-24)
$$(v^m v')' + h(v) = 0$$
$$v(0) = v(\lambda) = 0$$

this time the reaction term being a cubic of the kind studied in [31]:

(3-4-25) $$h(v) = -v(v - \alpha)(v - \beta) \quad , \quad 0 < \alpha < \beta.$$

We only look for positive solutions of course.

(3-4-24) can be written in the form of (3-4-1) with

$$f(u) = h(v) \quad , \quad u = D(v) = \frac{1}{m+1} v^{m+1}$$

so modulo a factor we have $f(u) = h(u^q)$ with some $q > 0$, and f looks as is shown in figure 3.4.4. Thus we can take $r = D(\alpha)$ and $r + a = D(\beta)$. As a necessary and sufficient condition for existence of positive solutions we have to assume (3-4-6) which follows for all $u \in [0, r + a[$ if we only require it for $u = 0$ and this condition becomes by a change of variables

(3-4-26)
$$\int_0^\beta v^m h(v)\, dv > 0.$$

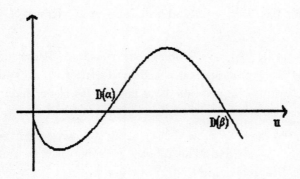

Figure 3.4.4

T is then defined on some interval $]0, b[$, and since $f(0) = 0$ we have to check the behaviour of T near 0 with proposition 3.4.2 (ii), (iii).

With $u = D(v)$, $dv/du = 1/d(v) = v^{-m}$ we get

(3-4-27)
$$f'(u) = h'(v)v^{-m}.$$

Thus for $m \leq 0$ (the increasing stickiness case) it follows that $f'(0) > -\infty$ and certainly (3-4-21) holds near $u = 0$. Hence

$$T(p) \to \infty \text{ as } p \to 0 \text{ for } m \leq 0.$$

If m is positive then $f'(0) = -\infty$ and the singularity is strong enough as to imply that $T(0)$ is finite:

$$f'(u) = h(v)v^{-m} = const\, h'(v)u^{-m/(m+1)} \leq -cu^{-m/(m+1)}$$

with some constant $c > 0$ since $h'(0) < 0$. Thus f' is growing faster near 0 than $(-\ln u)^q$ for any $q > 2$, so (3-4-20) is satisfied and we get

$$T(p) \to \lambda_0 \text{ as } p \to 0 \text{ with } 0 < \lambda_0 < \infty \text{ if } m > 0.$$

Next we are concerned with the shape of the positive solution branch. Because of proposition 3.4.2 (iv) it will always start to the left i.e. $T' < 0$ near 0. Also it will turn to the right afterwards since $f(r + a) = 0$ with f' being nonsingular, so $T(p) \to \infty$ as $p \to b$. In [12] it is proved that this is the only turn of the branch for any $m > -1$.

Without further effort we are only able to recover this result if we assume that $-1 < m \leq 1/2$:

Using theorem 3.4.4 we just have to show that f is an A-B-function on $]r, r+a[$. For $m < 0$ this is easy since then the Schwarzian of h as well as of D^{-1} is negative and we can use proposition 1.5.8 in order to show that $f = h \circ D^{-1}$ is an A-B-function. The case $m = 0$ is clear anyway.

Now let $m > 0$. We have to show that f is an A-function on the subinterval of $]r, r+a[$ where f' is positive. This follows immediately if we can show that f''' is negative there. Differentiating (3-4-27) we get

$$f'''(u) = v^{-3m-2}s(v, m)$$

with

$$s(v, m) := h'''(v)v^2 - 3mh''(v)v + m(2m+1)h'(v)$$

and we want to show

(3-4-28) $$s(v, m) < 0 \text{ for } 0 \leq m \leq 1/2 \text{ and } \alpha < v < \gamma$$

where $]\alpha, \gamma[$ is the subinterval of $]\alpha, \beta[$ on which h' is positive.

Now
$$\frac{\partial^2}{\partial m^2} s(v,m) = 4h'(v) > 0$$

for all $v \in {]}\alpha, \gamma{[}$. Thus for our claim it is sufficient to show that

(3-4-29) $s(v,0) < 0$ for all $v \in {]}\alpha, \gamma{[}$

and

(3-4-30) $s(v,1/2) < 0$ for all $v \in {]}\alpha, \gamma{[}$.

(3-4-29) is clear since $h''' < 0$, and for (3-4-30) it can be calculated directly that

$$s(v,1/2) = -(\alpha+\beta)v - \alpha\beta < 0.$$

As a result we have shown that T' does have exactly one zero which is a minimum of T. ∎

We finish this chapter with some remarks about Neumann time maps. Along the lines of section 1 in chapter 2 we can apply the results for Dirichlet time maps of section 3.1 to Neumann time maps. For the non bifurcating Neumann branches there is an analogue of theorem 3.4.3, but 3.4.4 cannot in any case be carried over to the Neumann problem since one has to introduce different definitions of $\hat{p}$ for $p > 0$ and $p < 0$. In the case that f is antisymmetric in a zero of f things work again. The interested reader can easily work out details for himself.

It should also be mentioned that we can apply our results to problems

$$u'' + h(u,\lambda) = 0 \qquad u'(0) = u'(1) = 0$$

where h might depend on λ differently from $h(u,\lambda)$ just being $\lambda^2 f(u)$. The framework for this is set up in [26], applications to cell aggregation models can be found in [27].

CHAPTER IV

GENERAL PROPERTIES OF TIME MAPS

4.1 Time maps and stability

We will discuss solutions of the Dirichlet problem

(4-1-1)
$$u'' + f(u) = 0$$
$$u(0) = u(\lambda) = 0$$

looking at all possible solution branches, central or noncentral, i.e. we allow f to have sign changes on its interval of definition. Since we do not want to care too much about the definition set of the time map in this paragraph we assume for f:

(4-1-2) $I \subset \mathbb{R}$ is an interval , $0 \in I$, $f : I \to \mathbb{R}$ continuously differentiable

With this

(4-1-3)
$$u'' + f(u) = 0$$
$$u(0) = 0 \quad u'(0) = p$$

has a unique solution $u = U(\cdot, p)$ for each initial condition p, and we define the Dirichlet time map as usual by

(4-1-4) $D(T) := \{p \in \mathbb{R} \mid U(\cdot, p) \not\equiv 0 \text{ and } U(t, p) = 0 \text{ for some } t > 0\}$

and

(4-1-5) $T(p) := \min\{t > 0 \mid U(t, p) = 0\}$.

$D(T)$ is generally not an interval as in chapter 1 since some of the orbits of $U(\cdot, p)$ can be homoclinic ones. Nevertheless one can show (see also [33])

4.1.1 PROPOSITION. *Let (4-1-2) hold for f. Then $D(T) \setminus \{0\}$ is open and T is continuously differentiable on $D(T) \setminus \{0\}$.*

PROOF: Let $p \in D(T) \setminus \{0\}$. Then $u = U(\cdot, p)$ is symmetric in $T(p)/2$ with $u(T(p)/2) =: q \neq 0$, $u'(T(p)/2) = 0$, $u''(T(p)/2) = -f(q) \neq 0$, all by some application of the existence an uniqueness theorem. Let us assume without loss of generality that $f(q) > 0$. Then q is a maximum of u and u' is positive on $[0, T(p)/2$ since otherwise u would have a positive minimum and thus would have to be a periodic solution without any zeroes. On $[0, T(p)/2[$ we have that u solves

$$(4\text{-}1\text{-}6) \qquad\qquad u' = \sqrt{p^2 - 2F(u)} \quad u(0) = 0$$

with F as usual being the integral of f with $F(0) = 0$. Thus

$$(4\text{-}1\text{-}7) \qquad F(q) = \frac{1}{2}p^2 \quad , \quad F(u) < F(q) \text{ for all } u \in [0, q[\quad , \quad f(q) > 0 .$$

Now let $|p - \hat{p}| < \epsilon$. Then since $f(q) > 0$ there is some $\hat{q}$ near q such that (4-1-7) holds for $\hat{p}$, $\hat{q}$ provided ϵ is small enough. Thus for $\hat{u}$ being the solution of (4-1-6) with p replaced by $\hat{p}$ it follows that the time t_0 which $\hat{u}$ needs to reach its maximum is

$$t_0 = \int_0^{\hat{q}} \frac{du}{\sqrt{\hat{p}^2 - 2F(u)}} < \infty .$$

By inflection we can continue $\hat{u}$ as a solution to (4-1-3) with initial condition $u'(0) = \hat{p}$, and thus $\hat{p} \in D(T)$.

To show that T is continuously differentiable on $D(T) \setminus \{0\}$ we just apply the implicit function theorem to

$$(4\text{-}1\text{-}8) \qquad\qquad \frac{\partial}{\partial t} U(t, p) = 0 .$$

If $p_0 \in D(T) \setminus \{0\}$ then $(T(p_0)/2, p_0)$ solves (4-1-8) with

$$\frac{\partial^2}{\partial t^2} U(T(p_0)/2, p_0) \neq 0 .$$

Since the left hand side of (4-1-8) is continuously differentiable there is a continuously differentiable curve $p \mapsto (t(p), p)$ defined in some neighbourhood of p_0 solving (4-1-8). Now we are almost done since it only remains to be shown that $t(p)$ is also the first zero of $\partial/\partial t\, U(\cdot, p)$ thus implying that $t(p) = T(p)/2$. This is certainly true for p close enough to $p_0 \neq 0$ since $\partial/\partial t\, U(\cdot, p_0)$ is bounded away from 0 on any closed subinterval of $[0, T(p_0)/2[$. $\blacksquare$

We are now going to prove the connection between the direction of a solution branch and the stability of its solutions, a fact which is well known but rather nice to prove with time maps. For more general statements which also contain information about the dimension of the unstable manifold see [4]. By stability we mean linearized stability:

Let $\mu_1 < \mu_2 < \dots$ be the sequence of (real and simple) eigenvalues of the linearization of (4-1-1) about a solution u , i.e.

$$v_i'' + f'(u)v_i + \mu_i v_i = 0$$
$$v_i(0) = v_i(\lambda) = 0$$

(4-1-9)

with $v_i \neq 0$ being an eigenfunction for μ_i (see [10]). We then call u stable if $\mu_1 > 0$, unstable if $\mu_1 < 0$.

An important fact is that by the Krein-Rutman theorem the eigenspace of the first eigenvalue μ_1 is spanned by a positive first eigenfunction v_1 (see [10] or [30]).

Let us first show that only positive or negative solutions have the chance to be stable:

4.1.2 PROPOSITION. *Let* (λ, u) *be a solution of (4-1-1) such that* u *changes sign on* $]0, \lambda[$.

Then u *is unstable.*

PROOF: If u changes sign in $]0, \lambda[$ then u' must change sign at least twice in $]0, \lambda[$, thus there is some subinterval $[t_1, t_2]$ of $]0, \lambda[$ such that

$$u'(t_1) = u'(t_2) = 0 \quad , \quad u'|_{]t_1, t_2[} > (<)0 .$$

Let us assume " > " without loss of generality. Let then

$$w(t) := u'(t) .$$

Then

$$w > 0 \text{ on }]t_1, t_2[\quad , \quad w(t_1) = w(t_2) = 0 ,$$

(4-1-10) $$w'' + f'(u)w = 0 \text{ and } w'(t_1) > 0 , w'(t_2) < 0$$

because of uniqueness. Multiplying (4-1-10) with the first eigenfunction $v_1 > 0$ and integrating by parts we get

$$w'(t_2)v_1(t_2) - w'(t_1)v_1(t_1) = -\mu_1 \int_{t_1}^{t_2} wv_1 .$$

From this it follows that $\mu_1 < 0$, thus u is unstable. ∎

For the stability of positive and negative solutions we then have

4.1.3 PROPOSITION. *Let (λ, u) be some solution of (4-1-1) such that $u = U(\cdot, p)$, $p \in D(T) \setminus \{0\}$ and $\lambda = T(p)$.*

Then

$$u \text{ is unstable if } pT'(p) < 0$$

and

$$u \text{ is stable if } pT'(p) > 0.$$

PROOF: First assume that $0 \neq p \in D(T)$, $u = U(\cdot, p$, $\lambda = T(p)$ and $pT'(p) < 0$. Differentiating the equation $U(T(p), p) = 0$ with respect to p gives

$$0 = \frac{\partial}{\partial t} U(T(p), p) T'(p) + \frac{\partial}{\partial p} U(T(p), p) = -pT'(p) + \frac{\partial}{\partial p} U(T(p), p).$$

If we now define

$$w(t) := \frac{\partial}{\partial p} U(t, p)$$

then

(4-1-11) $$w(0) = 0 \quad, \quad w'(0) = 1 \quad, \quad w(\lambda) = pT'(p) < 0$$

and

(4-1-12) $$w'' + f'(u)w = 0.$$

(4-1-11) and (4-1-12) imply that $\mu_1 < 0$: Let $]0, t_0[$ be the maximal interval on which w is positive. Then $t_0 < \lambda$, $w(t_0) = 0$, $w'(t_0) < 0$ and by multiplying (4-1-12) with $v_1 > 0$ and integrating over $]0, t_0[$ we get using integration by parts

$$w'(t_0)v_1(t_0) = \mu_1 \int_0^{t_0} w(t)v_1(t)\, dt.$$

Hence $\mu_1 < 0$ and u is unstable.

If $pT'(p) > 0$ then we get with the same definition of w that

(4-1-13) $$w(0) = 0 \quad, \quad w'(0) = 1 \quad, \quad w(\lambda) > 0.$$

If we can show that herefrom we get

(4-1-14) $$w > 0 \text{ on }]0, \lambda[$$

then we are done, for then multiplication of (4-1-12) with v_1 and integration over $]0, \lambda[$ gives

$$-w(\lambda)v_1'(\lambda) = \mu_1 \int_0^\lambda w(t)v_1(t)\, dt\,,$$

and since $v_1'(\lambda)$ has to be negative it follows that $\mu_1 > 0$.

So let us show (4-1-14). Differentiating

$$\frac{1}{2}\left(\frac{\partial}{\partial t}U(t,p)\right)^2 + 2F(U(t,p)) = \frac{1}{2}p^2$$

with respect to p we get with "$'$" denoting differentiation with respect to t

(4-1-15) $$u'w' + f(u)w = p\,.$$

If we now assume that w has a first zero $t_0 < \lambda$ then $w'(t_0) \leq 0$ and

$$u'(t_0)w'(t_0) = p \neq 0\,.$$

Thus $u'(t_0)$ has the opposite sign of $p = u'(0)$ so it must follow that $t_0 > T(p)/2$.
Since $w(\lambda) > 0$ there must be then a last zero $t_1 > t_0 > T(p)/2$ of w in $]0, \lambda[$.
Hence by (4-1-15)

$$u'(t_1)w'(t_1) = p$$

and $u'(t_1)$ must have the opposite sign of p since $t_1 > T(p)/2$. This would imply that $w'(t_1) < 0$, a contradiction to t_1 being the last zero of w and $w(\lambda)$ being positive.

So we have shown (4-1-14) and the proof is finished. ∎

There also is a strong relationship between the time map and the Liapounov functional of

(4-1-16) $$u_t = u_{xx} + f(u)$$
$$u(t,0) = u(t, \lambda) = 0$$

which is given by

$$t \mapsto \int_0^\lambda \frac{1}{2}(u_x(t,x))^2 - F(u(t,x))\, dx$$

and can be shown to be decreasing along solutions of (4-1-16).

Now (4-1-1) is the stationary equation of (4-1-16), t taking over the meaning of x in (4-1-16). Then we can consider the Liapounov functional of some solution u of (4-1-1) with $u = U(\cdot, p)$, $\lambda = T(p)$ to be a function of p and thus define

$$(4\text{-}1\text{-}17) \quad V(p) := \int_0^{T(p)} \frac{1}{2} \left(\frac{\partial}{\partial t} U(t, p) \right)^2 - F(U(t, p)) \, dt \quad \text{for } p \in D(T) \setminus \{0\}.$$

With this we have that V is decreasing as T increases and vice versa:

4.1.4 PROPOSITION. *Let* V *be defined by (4-1-17). Then* V *is continuously differentiable on* $D(T) \setminus \{0\}$ *with*

$$V'(p) = -\frac{1}{2} p^2 T'(p).$$

PROOF: That V is continuously differentiable is clear and its derivative can be calculated by

$$V'(p) = T'(p)\left(\frac{1}{2} \left(\frac{\partial}{\partial t} U(t, p)\right)^2 - F(U(T(p), p))\right)$$

$$+ \int_0^{T(p)} \frac{\partial}{\partial t} U(t, p) \frac{\partial}{\partial t} \frac{\partial}{\partial p} U(t, p) - f(U(t, p)) \frac{\partial}{\partial p} U(t, p) \, dt$$

$$= \frac{1}{2} p^2 T'(p) + \frac{\partial}{\partial t} U(T(p), p) \frac{\partial}{\partial p} U(T(p), p) - \frac{\partial}{\partial t} U(0, p) \frac{\partial}{\partial p} U(0, p)$$

$$- \int_0^{T(p)} \frac{\partial}{\partial p} U(t, p) \left(\frac{\partial^2}{\partial t^2} U(t, p) + f(U(t, p)) \right) \, dt$$

$$= -\frac{1}{2} p^2 T'(p)$$

where we have used integration by parts and the equalities (4-1-11) for $\partial/\partial p \, U(\cdot, p) = w$. ∎

$V(p)$ can be considered as the energy level of the stationary solution $u = U(\cdot, p)$. If we have two solutions u_1 and u_2 of (4-1-1) for the same λ with u_1 having higher energy than u_2 then there is the chance that there is some orbit of (4-1-16) which connects u_1 to u_2. We show in the next proposition that an unstable solution always has a higher energy than the next stable solution. In [4] it is shown that there exists an orbit connection between those two solutions.

4.1.5 PROPOSITION. Let $p_1 < p_2$ with $[p_1, p_2] \subset D(T) \setminus \{0\}$. *Further assume that*

$$T(p_1) = T(p_2) = \lambda \text{ and } T(p) \neq \lambda \text{ for all } p \in \,]p_1, p_2[\,.$$

Then

$$V(p_1) > V(p_2) \quad \text{if } p_1 T'(p_1) < 0$$

and

$$V(p_1) < V(p_2) \quad \text{if } p_1 T'(p_1) > 0\,.$$

PROOF: Without restriction we assume that $p_1 > 0$ and $T'(p_1) < 0$. It then follows that $T(p) < \lambda$ for all $p \in \,]p_1, p_2[\,$.

Also by a small perturbation of T which does not alter the sign of $V(p_1) - V(p_2)$ we can always achieve that critical points of T in $\,]p_1, p_2[\,$ are nondegenerate.

Let us first assume the case that T has a single minimum p_0 in $\,]p_1, p_2[\,$ with $T' < 0$ on $\,]p_1, p_0[\,$ and $T' > 0$ on $\,]p_0, p_2[\,$. For $E \in \,]T(p_0), \lambda]$ we can then define functions P_1 and P_2 implicitly by

$$p_1 \leq P_1(E) < p_0 \quad , \quad T(P_1(E)) = E \quad , \quad p_0 < P_2(E) \leq p_2 \quad , \quad T(P_2(E)) = E\,.$$

Then $P_i'(E) = 1/T'(P_i(E))$ and we get with proposition 4.1.4

$$\frac{d}{dE}(V(P_1(E)) - V(P_2(E))) = \frac{V'(P_1(E)))}{T'(P_1(E))} - \frac{V'(P_2(E))}{T'(P_2(E))}$$

$$= \frac{1}{2}\left((P_2(E))^2 - (P_1(E))^2\right) > 0\,.$$

Since $V(P_1(E))$ and $V(P_2(E))$ coincide at $E = T(p_0)$ the proof is finished in this case.

The general case that T has more critical points in $\,]p_1, p_2[\,$ we just prove by the following picture:

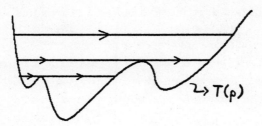

Figure 4.1.1

Arrows indicate directions in which $V(p)$ is decreasing. The arrows can be proved subsequently using the same method as in the first part of the proof. ∎

4.2 An inverse problem for positive solution branches

In this paragraph we will discuss the class of problems (4-1-1) where f ranges over all functions in $\mathcal{F}_a$, $a > 0$ (see (3-1-4)) which is a fairly general class of positive nonlinearities. Then we consider the operator τ which assigns to f its time map. We will give answers to the following questions:

(i) Is any curve one can draw a solution branch of (4-1-1)? Or: Which is the image of τ?

(ii) Is it possible to get the problem back from its positive solution branch? Or: Is τ invertible?

It is fairly clear that (ii) must have a positive answer if we look at the time map formula in proposition 3.1.1. In the same way as described in the beginning of section 3.3 we can derive that $T(\sqrt{x}) = (L\tilde{g})(x)$ where L is an integral operator with kernel $(x - y)^{-1/2}$ and $\tilde{g}(y) = g'(\sqrt{y})/\sqrt{y}$. Since (3-3-3) holds for L it should be possible to calculate back g from T and subsequently derive f back from g.

The operator which assigns g to f has always been called Φ. It can be defined on

(4-2-1)
$$\mathcal{F} := \bigcup \{\mathcal{F}_a \mid a \in \,]0, \infty]\}.$$

If $f \in \mathcal{F}_a$ then $g = \Phi(f) \in \mathcal{G}_b$ as defined in proposition 3.1.1. The image of Φ is thus contained in

(4-2-2)
$$\mathcal{G} := \bigcup \{\mathcal{G}_b \mid b \in \,]0, \infty]\}$$

and it is easy to show that

4.2.1 PROPOSITION.

$$\Phi : \mathcal{F} \longrightarrow \mathcal{G}$$

is a one to one map with $\Phi^{-1} = \Psi$ *given by*

$$\Psi : \mathcal{G} \longrightarrow \mathcal{F}$$

$$\Psi(g)(u) := \frac{d}{du}(\frac{1}{2}g^{-1}(u))^2 \, .$$

Part of the definition of τ is a linear integral operator B defined on $\mathcal{G}$ by

(4-2-3)
$$B(g)(p) := \sqrt{\frac{2}{\pi}} \int_0^p \frac{1}{\sqrt{p^2 - x^2}} g'(x)\, dx \,.$$

Using the time map formula of proposition 3.1.1 we have

(4-2-4)
$$\tau(f) = \sqrt{2\pi}(B \circ \Phi)(f)$$

for any $f \in \mathcal{F}$.

In order to answer questions (i), (ii) we have to characterize the image of $\mathcal{G}$ under B and to show that B is invertible on its image.

Before we write down the inverse of B let us let us collect some information on the image of B:

4.2.2 LEMMA. *Let* $g \in \mathcal{G}_b$, $0 < b \le \infty$, *then*

$$T(p) := B(g)(p)$$

is defined on $]0, b[$ *and*

(4-2-5)
$$T \in L^\infty_{loc}(]0, b[),$$

(4-2-6)
$$p \mapsto pT(p) \in L^1_{loc}([0, b[).$$

PROOF: In proposition 3.1.2 we have shown that T is locally Hölder continuous in $]0, b[$ thus (4-2-5) follows.

(4-2-6) holds because T is nonnegative and $pT(p)$ is the derivative of $\sqrt{2/\pi}S(p)$ with

$$S(p) := \int_0^p \sqrt{p^2 - x^2} g'(x)\, dx \,.$$

By lemma 4.2.2 the operator

(4-2-7)
$$C(T)(p) := \int_0^p qT(q)\, dq$$

is defined for $T = B(g)$, $g \in \mathcal{G}$. With this we can calculate g from $B(g)$:

4.2.3 LEMMA. For any $g \in \mathcal{G}_b$, $0 < b \leq \infty$ we have

(4-2-8)
$$(C \circ B)(g) \in \mathcal{G}_b,$$

(4-2-9)
$$(B \circ C \circ B)(g) = g$$

and

(4-2-10)
$$(C \circ B)(g) = (B \circ C)(g).$$

PROOF: (4-2-8) has been proved in lemma 4.2.2.

For a proof of (4-2-9) let $g \in \mathcal{G}_b$ and

$$T(p) := B(g)(p) = \sqrt{\frac{2}{\pi}} \int_0^p \frac{1}{\sqrt{p^2 - x^2}} g'(x) \, dx$$

for $p \in \,]0, b[$. Then by (4-2-8) $B \circ C \circ B)(g) = (B \circ C)(T)$ is defined on $]0, b[$, and

$$(B \circ C \circ B)(g)(x) = \sqrt{\frac{2}{\pi}} \int_0^x \frac{1}{\sqrt{x^2 - p^2}} p T(p) \, dp$$

$$= \frac{2}{\pi} \int_0^x \frac{p}{\sqrt{x^2 - p^2}} \int_0^p \frac{1}{\sqrt{p^2 - y^2}} g'(y) \, dy \, dp$$

$$= \frac{2}{\pi} \int_0^x g'(y) \int_x^y \frac{p}{\sqrt{x^2 - p^2}\sqrt{p^2 - y^2}} \, dp \, dy$$

by Fubini's theorem.

On the inner integral we use the transformation $p^2 = y^2 + t^2(x^2 - y^2)$, $p \, dp = t(x^2 - y^2) \, dt$ and get

$$\int_y^x \frac{p}{\sqrt{x^2 - p^2}\sqrt{p^2 - y^2}} \, dp = \int_0^1 \frac{dt}{\sqrt{1 - t^2}} = \frac{\pi}{2}.$$

Thus

$$(B \circ C \circ B)(g)(x) = \int_0^x g'(y) \, dy = g(x).$$

For a proof of (4-2-10) we observe that from the proof of (4-2-2) in lemma 4.1.2 it follows that

$$(C \circ B)(g)(p) = \int_0^p \sqrt{p^2 - x^2} g'(x) \, dx$$

$$= \int_0^p \frac{x}{\sqrt{p^2 - x^2}} g(x) \, dx = (B \circ C)(g)(p).$$

For the last but one equality we have used integration by parts and the fact that $g(0) = 0$. ∎

We are now ready to state the theorem of this paragraph:

4.2.4 THEOREM. *For* $0 < b \leq \infty$ *let*

$$\mathcal{T}_b := \{T :]0, b[\, \to \mathbb{R} \mid C(T) \text{ is defined with } C(T) \in \mathcal{G}_b, \, (B \circ C)(T) \in \mathcal{G}_b\}$$

and

$$\mathcal{T} := \bigcup\{\mathcal{T}_b \mid 0 < b \leq \infty\}.$$

Then

(i)

$$B : \mathcal{G}_b \longrightarrow \mathcal{T}_b$$

is invertible with

$$B^{-1} = B \circ C.$$

(ii) *The time map operator*

$$\tau : \mathcal{F} \longrightarrow \mathcal{T}$$

is defined by

$$\tau(f) = \sqrt{2\pi}(B \circ \Phi)(f)$$

and is invertible with

$$\tau^{-1}(T) = \Psi(\frac{1}{\sqrt{2\pi}}(B \circ C)(T)).$$

PROOF: Ad (i): By lemma 4.2.3 we have shown that $B(\mathcal{G}_b) \subset \mathcal{T}_b$ and that

$$(B \circ C) \circ B = id_{\mathcal{G}_b}.$$

In order to show that $B(\mathcal{G}_b) = \mathcal{T}_b$ let $T \in \mathcal{T}_b$. Then

$$g(x) := (B \circ C)(T)(x)$$

is in $\mathcal{G}_b$ and by lemma 4.1.3

$$(C \circ B)(g) = (B \circ C)(g) = (B \circ C \circ B \circ C)(T) = C(T)$$

since $C(T) \in \mathcal{G}_b$.

But since C is injective we also habe

$$B(g) = T.$$

With this

$$B(\mathcal{G}_b) = T_b$$

and

$$B \circ (B \circ C) = id_{T_b} .$$

We have thus shown (i).

(ii) is a consequence of this and proposition 4.2.1. ∎

The set $\mathcal{T}$ is now the set of all possible time maps of $f \in \mathcal{F}$. But at present the real information of what characterizes this set is buried under symbols. So let us reformulate the theorem in a somewhat less abstract language:

4.2.5 THEOREM. $T :]0, b[\rightarrow \mathbb{R}$ *is a time map of* $f \in \mathcal{F}$ *iff*

$$(4\text{-}2\text{-}11) \qquad g(x) = \frac{1}{\pi} \int_0^x \frac{1}{\sqrt{x^2 - p^2}} pT(p) \, dp = \frac{1}{\pi} \int_0^{\pi/2} (pT(p))|_p = x \sin \theta \, d\theta$$

is defined for $x \in]0, b[$ *and is a function in* $\mathcal{G}_b$, *i.e*

$$(4\text{-}2\text{-}12) \qquad g \in H^{1,1}_{loc}([0, b[) \quad , \quad g' \in L^\infty_{loc}(]0, b[) \quad , \quad g(0) = 0$$

and

$$(4\text{-}2\text{-}13) \qquad g'(x) > 0 \quad a.e. \ in \ ]0, b[.$$

With g *defined by (4-2-11) one can calculate* f *back from* T *via*

$$(4\text{-}2\text{-}14) \qquad a := \lim_{x \to b} g(x) \quad , \quad f(u) := \frac{d}{du} \left(\frac{1}{2} g^{-1}(u) \right)^2 \quad for \ u \in]0, a[.$$

The crucial condition for time maps is thus (4-2-13); (4-2-12) just describes the amount of regularity of T.

With this we can see immediately that T is a cone:

$$T_1, T_2 \in T \Rightarrow T_1 + T_2 \in T$$

(4-2-15) $\qquad T \in T, \alpha > 0 \Rightarrow \alpha T \in T$

$$T \in T \Rightarrow -T \notin T.$$

Hereby it is understood that the sum $T_1 + T_2$ is defined on the intersection of the definition sets of T_1 and T_2. T is certainly an open cone if we define the "right" topology on G and induce it upon T via B. But this does not give so much information, so we will not further persue this idea.

It is not so easy to tell the exact regularity of functions in T. We have seen in proposition 3.1.2 that a time map in T_b necessarily is in $C_{loc}^{0+1/2}$. This might not be enough regularity, but if we assume a little bit more, namely $C_{loc}^{0+\alpha}$ - regularity with $\alpha > 1/2$ then the derivative of g exists:

4.2.6 PROPOSITION. *Let $T \in C_{loc}^{0+\alpha}(]0, b[)$ with $\alpha > 1/2$ and assume that T is positive on $]0, b[$ with $T \in L_{loc}^1([0, b[)$.*

Further assume that the map

$$p \mapsto pT(p)$$

is strictly increasing on $]0, b[$.

Then $T \in T_b$.

PROOF: By theorem 4.2.5 we have to show that (4-2-12) , (4-2-13) holds for the map g defined by (4-2-11).

Since T is integrable near 0 and positive and since $pT(p)$ is increasing it has to follow that $pT(p) \to 0$ as $p \to 0$. Thus $p \mapsto pT(p)$ is continuous, g is defined and $g(0) = 0$.

Let us show that g is differentiable:

By (4-2-11) we have for $x \in]0, b[$

$$\pi g(x) = \int_0^{\pi/2} x \sin \theta T(x \sin \theta) \, d\theta$$

$$= -\int_0^{\pi/2} \frac{\sin \theta}{\cos \theta} \frac{d}{d\theta} \left(\int_{x \sin \theta}^x T(p) \, dp \right) d\theta$$

$$= \int_0^{\pi/2} \frac{1}{\cos^2 \theta} \int_{x \sin \theta}^x T(p) \, dp \, d\theta \, .$$

Integration by parts is allowed since we can show the last integrand to be continuous in $\theta = \pi/2$ using de l'Hôpital's theorem. If we denote this integrand by $i(x, \theta)$ then i has a derivative with respect to x which is continuous on $[0, \pi/2[$:

$$\frac{\partial}{\partial x} i(x, \theta) = \frac{T(x) - \sin\theta\, T(x \sin\theta)}{\cos^2\theta} = \frac{1 - \sin\theta}{\cos^2\theta} T(x) + \sin\theta \frac{T(x) - T(x\sin\theta)}{\cos^2\theta}.$$

The first summand of the right hand side term is continuous on $[0, \pi/2]$. So it suffices to estimate the second summand for θ near $\pi/2$ by some integrable function.

Since T is locally $C^{0+\alpha}$ there is some constant $c > 0$ such that for all θ near $\pi/2$

$$\frac{|T(x) - T(x\sin\theta)|}{\cos^2\theta} \le c \frac{x^\alpha (1 - \sin\theta)^\alpha}{\cos^2\theta}$$

$$\le c x^\alpha \left(\frac{1 - \sin\theta}{\cos^2\theta}\right)^\alpha (\cos\theta)^{-2(1-\alpha)} \le \tilde{c}(\cos\theta)^{-2(1-\alpha)}$$

with some constant $\tilde{c}$ independent of θ and exponent $-2(1 - \alpha) > -1$ by assumption.

Thus we have shown that g' exists with

(4-2-16)
$$\pi g'(x) = \int_0^{\pi/2} \frac{1}{\cos^2\theta} (T(x) - \sin\theta\, T(x \sin\theta))\, d\theta$$
$$= \int_0^x x\left(x^2 - p^2\right)^{-3/2} (xT(x) - pT(p))\, dp.$$

With the same proof we have shown that g' is locally bounded. From the assumption that $pT(p)$ is strictly increasing it follows with (4-2-16) that g' is positive. ∎

4.3 Applications and examples

As a first application we can prove generic properties of time maps and get a result analogous to [3], [33]:

We show that the set of functions which have time maps with nondegenerate critical points is in a certain sense dense in $\mathcal{F}$. A critical point of T is a point p_0 with $T'(p_0) = 0$; it is nondegenerate if $T''(p_0) \neq 0$. A critical point of T corresponds to a solution of (4-1-1) for which the linearized problem has a zero eigenvalue. The result thus shows that in a subset of the (λ, u)-space which is bounded in u there are for generic f at most finitely many solutions of (4-1-1) which are nonhyperbolic, i.e. whose linerization has a zero eigenvalue.

We have to specify the meaning of generic and therefor introduce the topology of local L^1-convergence on $\mathcal{F}_a$:

$$f = \lim_{n \to \infty} f_n \text{ in } \mathcal{F}_a \quad \text{iff} \quad f = \lim_{n \to \infty} f_n \text{ in } L^1([0, c]) \text{ for all } 0 < c < a.$$

4.3.1 THEOREM. *For any $a > 0$ the set of $f \in \mathcal{F}_a$ which have a C^∞-time-map with only nondegenerate critical points is a dense subset of $\mathcal{F}_a$.*

PROOF: Let $f \in \mathcal{F}_a$. Since the set of C^∞-functions in $\mathcal{F}_a$ is dense in $\mathcal{F}_a$ we can assume without loss of generality that $f \in C^\infty(]0, a[)$. Then $g := \Phi(f)$ is in $C^\infty(]0, b[)$ with $b = \sqrt{2F(a)}$, as well as $T = \tau(f) = \sqrt{2\pi}B(g)$, the time map of f.

We now want to add a function $\delta H(p)$ to $T(p)$, δ small, such that the result is still a time map but with all critical points being nondegenerate.

For this we take some function $H \in C^\infty([0, b[)$ with the following properties:

(4-3-1)
$$\frac{d}{dp}(pH(p)) > 0,$$
$$H'(p) \neq 0 \text{ for all } p \in]0, b[,$$
$$pH(p) \leq c < \infty \text{ for all } p \in]0, b[.$$

$H(p) := 1/(1 + p)$ would be a possible choice in any case, but we want to point out that there is a certain amount of freedom in the choice of the perturbation. For $b = \infty$ the above H guaranties that the perturbed time map has the same behaviour as T near ∞. For $b < \infty$ and $T(0) = 0$ we can choose a perturbation

H with $H(0) = 0$, $H' > 0$, and as a result the perturbed time map has initial value 0 too.

Now choose $\delta_0 > 0$. The function $h(p) := T'(p)/H'(p)$ is in $C^\infty(]0, b[$ by assumption. By Sard's lemma its set of critical values (values $h(p)$ with $h'(p) = 0$) is a set of measure zero. Thus there is some δ, $0 < \delta < \delta_0$ such that

$$h(p) = -\delta \Longrightarrow 0 \neq h'(p) = \frac{T''(p)H'(p) - T'(p)H''(p)}{(H'(p))^2} = \frac{1}{H'(p)}(T''(p) + \delta H''(p)).$$

So if we define

$$\widehat{T}(p) := A(T(p) + \delta H(p)) \quad, \quad p \in]0, b[\quad, \quad A > 0$$

then we have

$$\widehat{T}'(p) = 0 \Longrightarrow \widehat{T}''(p) \neq 0,$$

critical points of $\widehat{T}$ are nondegenerate.

Thus it remains to be shown that $\widehat{T} \in \mathcal{T}_b$, i.e. by theorem 4.2.5 we have to show that $\hat{g}' > 0$ with

$$\hat{g}(x) := \frac{1}{\pi} \int_0^{\pi/2} (p\widehat{T}(p))|_p = x \sin\theta \, d\theta.$$

If we define

$$\tilde{g}(x) := \frac{1}{\pi} \int_0^{\pi/2} (pH(p))|_p = x \sin\theta \, d\theta$$

then we get by (4-3-1) that

$$\tilde{g}' > 0 \text{ on }]0, b[\quad \text{and} \quad \tilde{g}(b) < \infty.$$

Hence

$$\hat{g}' = A(g' + \tilde{g}') > Ag' > 0 \text{ on }]0, b[$$

and $\widehat{T} \in \mathcal{T}_b$.

Calculating $\hat{f} := \tau^{-1}(\widehat{T})$ we adjust A such that $\hat{f}$ is defined on $]0, a[$: By (4-2-14) $\hat{f}$ is defined on $]0, \hat{g}(b)[$ with

$$\hat{g}(b) = A(g(b) + \delta\tilde{g}(b)) = \begin{cases} A(a + \delta\tilde{g}(b)) < \infty \text{ if } a < \infty \\ \infty \\ , \textit{if } a = \infty. \end{cases}$$

So if we define

$$A := \begin{cases} \dfrac{a}{a + \delta \tilde{g}(b)} & \text{for } a < \infty \\[2mm] 1 & \text{for } a = \infty \end{cases}$$

then we get that $\hat{f} \in \mathcal{F}_a$ and $\hat{f} = \Psi(\hat{g})$ is certainly arbitrarily close to f in L^1_{loc} is δ_0 is small enough. ∎

One can use the same method of proof in order to show similar results if we take subsets of $\mathcal{F}_a$ instead of $\mathcal{F}_a$ such as the set of all C^k-f with $f(0) = 0$ ($f(0) > 0$) or the set of all C^k-f with $f(a) = 0$ or intersections, hereby also choosing a different kind of topology. Since $\hat{f}(0)$, $\hat{f}(a)$ etc. can be calculated from $\hat{g}$ and thus from $\hat{T}$ we just have to make the right choice of H such that $\hat{f}$ belongs to the same subset as f. Also note that the requirements on H need not be as strong as in (4-3-1): For $\hat{T}$ to be a time map it suffices that H is a time map. For the nondegeneracy condition we can allow H to have finitely many nondegenerate critical points and we can still adjust δ such that critical points of $\hat{T}$ are nondegenerate.

We next use the time map formula

(4-3-2)
$$T(p) = 2 \int_0^{\pi/2} g'(p \sin \theta) \, d\theta$$

and its inverse

(4-3-3)
$$\frac{g(x)}{x} = \frac{1}{\pi} \int_0^{\pi/2} T(x \sin \theta) \sin \theta \, d\theta$$

in order to establish some connections between properties of f and T. As a first result we get λ-regions of existence and nonexistence for (4-1-1):

Using (4-3-2) we have

$$T(p) - \lambda = 2 \int_0^{\pi/2} g'(p \sin \theta) - \frac{\lambda}{\pi} \, d\theta .$$

Thus if $\pi g'(x) > (<)\lambda$ for all x then $T(p) > (<)\lambda$ for all p. With $u = g(x)$ we have $g'(x) = \sqrt{2F(u)}/f(u)$. As a result λ is not in the range of T if λ is not in the range of $u \mapsto \sqrt{2F(u)}/f(u)$.

In a similar way we get from (4-3-3) that

$$\frac{1}{\pi} \int_0^{\pi/2} T(x \sin \theta) - \lambda \, d\theta = \frac{g(x)}{x} - \frac{\lambda}{\pi}.$$

Hence if $T(p) > (<)\lambda$ for all p then $\pi g(x)/x > (<)\lambda$ for all x. With $u = g(x)$ we get $g(x)/x = u/\sqrt{2F(u)}$ and thus that λ is in the range of T if λ is in the range of $u \mapsto \pi u/\sqrt{2F(u)}$.

Trivial though the proof may be we find it worth while formulating a proposition for this result:

4.3.2 PROPOSITION. *If λ is not in the range of*

$$u \mapsto \pi \frac{\sqrt{2F(u)}}{f(u)} \quad , \quad u \in \,]0, a[$$

then (4-1-1) does not have any nontrivial solutions for this λ.

If λ is in the range of

$$u \mapsto \pi \frac{u}{\sqrt{2F(u)}} \quad , \quad u \in \,]0, a[$$

then there exists a nontrivial solution of (4-1-1) for this λ.

One can substitute the nonexistence condition by one which only involves knowledge of f':

Let us look for a sufficient condition such that $\lambda < \pi \sqrt{2F(u)}/f(u)$, i.e.

$$(4\text{-}3\text{-}4) \qquad\qquad 2F(u) - \frac{\lambda^2}{\pi^2} f^2(u) > 0$$

for all u. This is only possible for $f(0) = 0$. In this case we calculate the derivative of (4-3-4) which is

$$2f(u)(1 - \frac{\lambda^2}{\pi^2} f'(u)).$$

So (4-3-4) holds if

$$(4\text{-}3\text{-}5) \qquad\qquad f'(u) < \frac{\pi^2}{\lambda^2} \quad \text{for all } u \in [0, a[.$$

On the other hand we get $\lambda > \pi \sqrt{2F(u)}/f(u)$ for all u if

(4-3-6) $$f'(u) > \frac{\pi^2}{\lambda^2} \text{ for all } u \in [0, a[\,.$$

One can play more games with (4-3-2) and (4-3-3), e.g.

$$T(p) - \alpha p = 2 \int_0^{\pi/2} g'(p \sin \theta) - \frac{\alpha}{2} p \sin \theta \, d\theta\,.$$

Since $g'(x) - \alpha/2\, x = x(1/f(u) - \alpha/2)$ with $u = g(x)$ we get

(4-3-7)
$$f(u) < \frac{2}{\alpha} \text{ for all } u \Rightarrow T(p) > \alpha p \text{ for all } p\,,$$
$$f(u) > \frac{2}{\alpha} \text{ for all } u \Rightarrow T(p) < \alpha p \text{ for all } p\,.$$

Provided f, g, T have enough regularity we get differentiating (4-3-2) and (4-3-3)

(4-3-8) $$T'(p) = 2 \int_0^{\pi/2} g''(p \sin \theta) \sin \theta \, d\theta$$

and

(4-3-9) $$\frac{d}{dx}\left(\frac{g(x)}{x}\right) = \frac{1}{\pi} \int_0^{\pi/2} T'(x \sin \theta) \sin^2 \theta \, d\theta\,.$$

Using these we get the following relations between properties of f and T (the first two can be proved in the same way as theorem 1.3.2):

$$f \text{ sublinear} \Rightarrow T' > 0\,,$$
$$f \text{ superlinear} \Rightarrow T' < 0\,,$$
$$f' > 0 \Rightarrow T \text{ sublinear}\,,$$
$$f' < 0 \Rightarrow T \text{ superlinear}\,,$$
(4-3-10)
$$T' > 0 \Rightarrow \sqrt{F} \text{ sublinear}\,,$$
$$T' < 0 \Rightarrow \sqrt{F} \text{ superlinear}\,,$$
$$T \text{ sublinear} \Rightarrow F \text{ superlinear}\,,$$
$$T \text{ superlinear} \Rightarrow F \text{ sublinear}\,.$$

From the fifth and sixth of these assertions if follows that multiple solutions of (4-1-1) exist if $d/du\, u/\sqrt{F(u)}$ changes sign.

But let us persue the road of l'art pour l'art a little further since now that we have a characterization of the set of all possible solution branches it might be interesting to see which curves can be among them and which ones not.

With a first example we will show that there are solution branches having a whole interval of bifurcation points at 0 as well as at ∞:

Let $h : \mathbb{R} \to \mathbb{R}$ be a periodic function which is continuously differentiable. Then

$$T : \,]0, \infty[\, \to \mathbb{R} \quad , \quad T(p) := h(\ln p) + A$$

is a time map if $A > 0$ is large enough, for

$$\frac{d}{dp}(pT(p)) = h(\ln p) + h'(\ln p) + A > 0$$

for A large enough, since h and h' are bounded. Thus by proposition 4.2.6 T is the solution branch of some problem (4-1-1) with $f \in \mathcal{F}$.

It is then of interest how f has to look like to give a solution branch like this. In this case we can give a complete answer since

$$\frac{f(u)}{u} = h_1(\ln u) \quad , \quad h_1 \text{ periodic with period } P$$

$$\text{iff} \quad T(p) = h_2(\ln p) \quad , \quad h_2 \text{ periodic with period } P \,.$$

The if-part we can prove using (4-3-3), for then it follows that $g(xe^P)/(xe^P) = g(x)/x$, thus

$$g(xe^P) = e^P g(x)$$

and

$$F(e^P g(x)) = F(g(e^P x)) = \frac{1}{2}x^2 e^{2P} = e^{2P} F(g(x)) \,.$$

From this it follows with $u = g(x)$ and differentiation that

$$f(e^P u) = e^P f(u) \,,$$

thus f has to have a representation $f(u) = u h_1(\ln u)$, h_1 being periodic with period P.

The only-if-part follows directly from the initial value problem since we can show that $U(\cdot, e^{P+q}) = e^P U(\cdot, e^q)$.

In the first example we have used that if

$$T(p) = h(\ln p)$$

then

$$\frac{d}{dp}(pT(p)) > 0 \quad \text{iff} \quad h(s) + h'(s) > 0.$$

We can put weaker assumptions on h if we e.g. make an Ansatz

$$T(p) = h(\ln(1 - \ln p)) \quad , \quad 0 < p < 1.$$

Then

$$\frac{d}{dp}(pT(p)) > 0 \quad \text{iff} \quad h(s) - e^{-s}h'(s) > 0, \, s \in]0, \infty[.$$

If we can show that this is the case for e.g.

$$h(s) = s(\sin s + 1) + \frac{A}{s}$$

with $A > 0$ large enough then we have an example of a time map having all the interval $[0, \infty[$ as a limit set as $p \to 0$. This is because $s := \ln(1 - \ln p) \to \infty$ as $p \to 0$, and h approaches $[0, \infty[$ as $s \to \infty$. Now

$$h(s) - e^{-s}h'(s) = s(\sin s + 1) + \frac{A}{s} - e^{-s}(\sin s + 1) - se^{-s}\cos s + \frac{Ae^{-s}}{s^2}$$

$$> \frac{A}{s} - 2e^{-s} - se^{-s} = \frac{1}{s}\left(A - e^{-s}(2s + s^2)\right) \geq \frac{c}{s}$$

for some $c > 0$ if we choose A large enough.

The corresponding f has to oscillate fairly wildly near 0 though it still follows that $f(u) = x/g'(x) \to 0$ as $u = g(x) \to 0$. From proposition 4.3.2 it follows that $\sqrt{F(u)}/f(u)$ has to have the whole interval $[0, \infty[$ as a limit set as $u \to 0$.

Next we will turn to the non-time-maps, so to get an idea of what properties solution branches of (4-1-1) cannot have. If T is regular enough then we can calculate g' by (4-2-16) and get a necessary condition for T to be a time map:

$$\pi g'(x) = \int_0^x x \left(x^2 - p^2\right)^{-3/2} (xT(x) - pT(p)) \, dp.$$

Let $0 < \bar{x} < x$ be such that $pT(p) \leq xT(x)$ for $0 \leq p \leq \bar{x}$ and $pT(p) \geq xT(x)$ for $\bar{x} < p \leq x$. Then, since $d/dp\,(x^2 - p^2)^{-3/2} > 0$

$$\pi g'(x) \leq x \left(x^2 - \bar{x}^2\right)^{-3/2} \int_0^x xT(x) - pT(p)\,dp.$$

Hence a necessary condition for $T \in \mathcal{T}$, i.e. $g' > 0$ is

(4-3-11) $$\int_0^{\bar{x}} xT(x) - pT(p)\,dp > \int_{\bar{x}}^x pT(p) - xT(x)\,dp$$

for any such $\bar{x}$. In figure 4.3.1 the area A (A') has to be larger than B (B'):

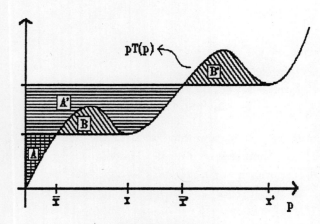

Figure 4.3.1

With this we immediately get that

$$T(p) = (p - a)^2 + b$$

with $a, b > 0$ is not a time map for b small. Minima of T either have to be "big enough" or else have to occur "late enough".

A question related to our first example is whether or not it is possible to have periodic time maps. The answer is no, unless $T \equiv const$:

Let

$$T(p) = h(p) + A$$

with $A > 0$, h periodic with period $P > 0$ and h having zero mean over one period. Let $h \not\equiv 0$ and (say) C^1. Then with (4-2-16)

$$\pi x g'(x) = \int_0^x \frac{x^2}{(x^2 - p^2)^{3/2}} (x T(x) - p T(x) + p(T(x) - T(p))) \, dp$$
$$= T(x) \int_0^x \frac{x^2}{(x^2 - p^2)^{3/2}} (x - p) \, dp + x^2 \int_0^x \frac{p}{(x^2 - p^2)^{3/2}} (T(x) - T(p)) \, dp.$$

Via integration by parts the first integral is equal to x, so

$$\pi g'(x) = T(x) + x I(x)$$

with

$$I(x) = \int_0^x \frac{p}{(x^2 - p^2)^{3/2}} (h(x) - h(p)) \, dp.$$

We will next define a sequence $x_n \to \infty$ for which $x_n I(x_n) \to -\infty$. Since T is bounded this will imply that $g'(x_n) \to -\infty$ and T cannot be a time map.

Let $h(x_0)$ be the minimum of h on $]0, P[$, and define

$$x_n := x_0 + nP.$$

Then

(4-3-12) $$h(x_n) < 0 \quad , \quad h(x_n) - h(p) \le 0 \text{ for all } p.$$

Next let $y_0 < x_0$ be the first zero of h which has to exist since h has a positive maximum. We assume without loss of generality that zeroes of h are isolated and simple. Then there exists another zero $z_0 < y_0$ of h with

$$h(z_0) = h(y_0) = 0, \; h|_{]z_0, y_0[} > 0, \; h|_{]y_0, x_0[} < 0.$$

Define

$$z_n := z_0 + nP \quad , \quad y_n := y_0 + nP.$$

Then because of (4-3-12)

$$I(x_n) \le \int_{z_n}^{y_n} \frac{p}{(x_n^2 - p^2)^{3/2}} (h(x_n) - h(p)) \, dp \le h(x_n) \int_{z_n}^{y_n} \frac{p}{(x_n^2 - p^2)^{3/2}} \, dp$$
$$= h(x_0) \left((x_n^2 - y_n^2)^{-1/2} - (x_n^2 - z_n^2)^{-1/2} \right).$$

$x_n^2 - y_n^2 = x_0^2 + 2nP(x_0 - y_0)$, with a corresponding formula holding for $x_n^2 - z_n^2$. Thus

$$\sqrt{x_n} I(x_n) \leq h(x_0) \sqrt{\frac{x_n}{n}} \left(\left(\frac{x_n^2 - y_n^2}{n} \right)^{-1/2} - \left(\frac{x_n^2 - z_n^2}{n} \right)^{-1/2} \right)$$

$$\longrightarrow h(x_0) \sqrt{P} \left((2P(x_0 - y_0))^{-1/2} - (2P(x_0 - z_0))^{-1/2} \right) < 0$$

for $n \to \infty$ which implies that $x_n I(x_n) \to -\infty$.

So there cannot be any periodic time maps. Looking at the foregoing proof one gets the idea that time maps can only oscillate with a fixed amplitude if oscillations become more and more slow as p increases. If a time map oscillates with a fixed frequency then the amplitude has to decrease as p grows. See also the next figures.

We finish with some pictures which show a comparison between the graphs of $u \mapsto u/f(u)$, $u \mapsto u/\sqrt{2F(u)}$ and $u \mapsto \sqrt{2F(u)}/f(u)$. From formulas (4-3-2) and (4-3-3) one could get the idea that T is at least as "wild" as $g(x)/x = u/\sqrt{2F(u)}$ and as most as "wild" as $g'(x) = \sqrt{2F(u)}/f(u)$ with $u/f(u)$ describing some state in between. If "wildness" is defined by range of values then this is true by proposition 4.3.2. Other connections are stated in (4-3-10). It might be interesting to look for other definitions of "wildness" which make the statement true. "Number of critical points" is not a right choice as we have seen in the beginning of section 3.3.

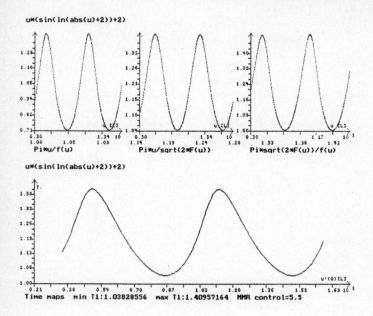

Figure 4.3.2

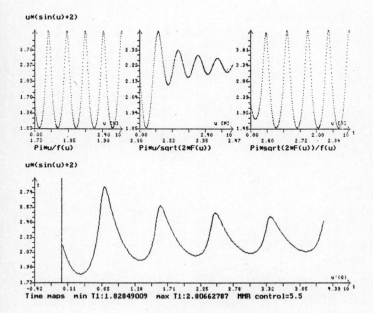

Figure 4.3.3

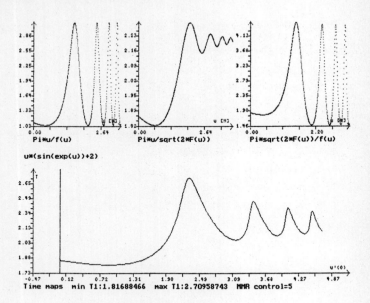

Figure 4.3.4

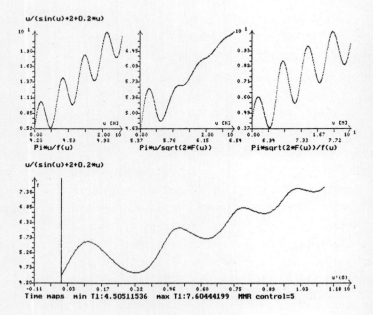

Figure 4.3.5

Appendix: Some remarks on the computer plots of time maps

The pictures of time maps we have shown so far were calculated by a program written in Pascal designed to run on an XT- or AT-compatible PC in reasonable time and with reasonable accuracy. The numerical scheme used is a modified midpoint rule (see [33]). We solve the initial value problem

(A-1-1)
$$u' = v \qquad v' = -f(u)$$
$$u(0) = 0 \qquad u'(0) = p$$

for p in $D(T)$ or some subinterval thereof. We have to calculate the first time t_1 for which $v(t_1) = 0$, then $T(p) = 2t_1$. Numerically it is not possible to hit t_1 exactly, so we solve (A-1-1) until

$$v(t_1 + \Delta t) > 0 \quad \text{if } p < 0$$
$$v(t_1 + \Delta t) < 0 \quad \text{if } p > 0$$

for the first mesh point $t_1 + \Delta t$. For an approximation of Δt we use

$$v(t_1 + \Delta t) \sim v(t_1) + \Delta t v'(t_1 + \Delta t) - \frac{1}{2}(\Delta t)^2 v''(t_1 + \Delta t)$$
$$= -\Delta t f(u(t_1 + \Delta t)) + \frac{1}{2}(\Delta t)^2 f'(u(t_1 + \Delta t))v(t_1 + \Delta t).$$

With this we get an approximation of t_1.

It is obvious that using a fixed stepsize h for the method is not useful, since $T(p)$ will generally have a large range of values, and with a fixed stepsize either the method is too crude for small $T(p)$ or takes too much time for large $T(p)$. Also our kind of problems often run into parameter ranges where interior or boundary layers occur, so the stepsize should vary along orbits so as to take proper account of these phenomena. We use the iteration process in our implicit method as a means to control h: The iteration is stopped when the relative error is less then $10^{-control}$, where $control$ is a parameter to be entered into the program by the user. Subsequently h is altered in a way as to ensure that the mean number of iterations is always 2. To make the process smooth we define "number of iterations" by the actual number of steps minus 1 plus the quotient of the error and $10^{-control}$. This method seems to work well even in tough situations.

As for the p-interval for which the time map should be calculated we find it better to enter the corresponding u-interval of intersections of orbits with the $v = 0$-axis in the phase plane. Having a definition of f one can define interesting parts of the phase plane more easily by this interval. In the program the u-interval is called $[umin, umax]$. The program the calculates the corresponding p-interval by first determining the set of all u with

$$F(u) - F(s) > 0 \quad \text{for all } 0 \leq s \leq u.$$

Those are all values u for which an orbit through $(u, 0)$ can reach the $u = 0$-axis. The corresponding p-values are given by $p = \pm\sqrt{2F(u)}$ according to the sign of u. The program now takes $pmin$ as the smallest of these p-values, $pmax$ as the largest and $[pmin, pmax]$ as the definition set of T. This way there can be some initial values of homoclinic orbits in $[pmin, pmax]$, but we trust in the fact that those orbits are numerically highly unstable.

In the case that $pmin < 0 < pmax$ we choose p-meshpoints such that $-p$ is in the mesh if p is and $-p \in [pmin, pmax]$. This way all time maps T_i can be calculated from T using the formulas

$$T_{2i}(p) = i(T(p) + T(-p)) \quad T_{2i+1}(p) = (i+1)T(p) + iT(-p).$$

For anybody who would like to use this program a discette is available from the author. Just indicate which type of discettes you are using. The software comes with a user friendly surface and supports CGA-, Hercules-, EGA- and VGA-graphics as well as truly compatibles.

REFERENCES

1. L.V.Ahlfors, "Compex Analysis," McGraw-Hill, Tokyo, 1966.

2. M.Braun, "Differential Equations and Their Applications," Springer Applied Mathematical Sciences vol. 15, Heidelberg - Berlin - New York, 1975.

3. P.Brunovský, S.N.Chow, *Generic properties of stationary steady state solutions of reaction diffusion equations*, J.Diff.Equ. **53** (1984), 1–23.

4. P.Brunovský, B.Fiedler, *Connecting orbits in scalar reaction diffusion equations. II. The complete solution*, J.Diff.Equ. **81** (1989), 106–135.

5. N.Chaffee, E.F.Infante, *A bifurcation problem for a nonlinear differential equation*, Applicable Anal. **4** (1974), 17–37.

6. M.Chipot, F.B.Weissler, *Blow up results for a nonlinear parabolic equation with a gradient term*, Univ. of Minnesota, IMA Preprint Series **298** (1987),.

7. S.N.Chow, J.K.Hale, "Methods of Bifurcation Theory," Springer, Berlin - Heidelberg - New York, 1982.

8. S.N.Chow, J.A.Sanders, *On the number of critical points of the period.*, J.Diff.Equ. **64** (1986), 51–66.

9. S.N.Chow, D.Wang, *On the monotonicity of the period function of some second order equations*, Casopis pre pestovani matematiky.

10. E.A.Coddington, N. Levinson, "Theory of Ordinary Differential Equations," McGraw Hill, New York, Toronto, London, 1955.

11. P.Collet, J.-P.Eckmann, "Iterated maps on the interval as dynamical systems," Birkhäuser Boston, 1980.

12. D.G.Aronson, M.G.Crandall, L.A.Peletier, *Stabilization of solutions of a degenerate nonlinear diffusion problem*, Nonlinear Anal. 6 (1982), 1001–1022.

13. M.G.Crandall, P.H.Rabinowitz, *Bifurcation from simple eigenvalues*, J.Functional Analysis **8** (1981), 321–340.

14. P.C.Fife, "Mathematical Aspects of Reacting and Diffusing Systems," Springer Lecture Notes Biomath. vol. 28, Heidelberg - Berlin - New York, 1979.

15. D.A.Frank-Kamenskii, "Diffusion and Heat Transfer in Chemical Kinetics," Plenum Press, New York - London, 1969.

16. S.Heintze, *Travelling waves for semilinear parabolic partial differential equations in cylindrical domains*, Univ. Heidelberg, SFB 123 Preprint Series **506** (1989).

17. D.Henry, "Geometric Theory of Semilinear Parabolic Equations," Springer Lecture Notes in Mathematics vol. 840, Heidelberg-Berlin-New York, 1981.

18. J.L.Kaplan, J.A.Yorke, *Ordinary differential equations which yield periodic solutions of differential delay equations*, J.Math.Anan.Appl. 48 (1974), 317–324.

19. S.Karlin, "Total Positivity," Stanford University Press, Stanford, California, 1968.

20. B.Kawohl, L.A.Peletier, *Observations on blow up and dead cores for nonlinear parabolic equations*, Univ. Heidelberg, SFB 123, Preprint Series **481** (1988),.

21. R.Nussbaum, *Uniqueness and nonuniqueness for periodic solutions of $x'(t) = -g(x(t-1))$*, J.Diff.Equ. **34** (1979), 24–54.

22. Z.Opial, *Sur les périodes des solutions de l'équation differentielle* $x"+g(x) = 0$, Ann.Polon.Math. **10** (1961), 49–72.

23. H.O.Peitgen, K.Schmitt, *Global analysis of two-parameter elliptic eigenvalue problems*, Trans.Amer.Math.Soc. **283** (1984), 57–95.

24. P.Rabinowitz, *Some global results for nonlinear eigenvalue problems*, J.Functional Analysis **7** (1971), 487–513.

25. F.Rothe, *The periods of the Volterra-Lotka-systems*, J.Reine Angew. Math. **355** (1985), 129–138.

26. R.Schaaf, *Global behaviour of solutions branches for some Neumann problems depending on one or several parameters*, J.Reine Angew. Math. **346** (1984), 1–31.

27. R.Schaaf, *Stationary solutions of chemotaxis systems*, Trans.AMS **292** (1985), 531–556.

28. P.Schuster,K.Sigmund, *Coyness, philandering and stable strategies*, Animal Behaviour **29** (1981), 186–192.

29. K.Schmitt, H.L.Smith, *Eigenvalue problems for nondifferentiable mappings*, J.Diff.Equ. **33** (1979), 294–319.

30. J.Smoller, "Shock Waves and Reaction/Diffusion equations," Springer Grundlehren vol. 258, Heidelberg - Berlin - New York, 1983.

31. J.Smoller, A.Wassermann, *Global bifurcation of steady state solutions*, J.Diff.Equ. **39** (1981), 269–290.

32. J.Smoller, A. Wassermann, *Symmetry-breaking for positive solutions of semilinear elliptic equations.*, Arch. Rat. Mech. Anal. **95** (1986), p. 217–225.

33. J.Smoller, A. Wassermann, *Generic bifurcation of steady state solutions*, J.Diff.Equ. **39** (1981), 432–438.

34. D.Stoffer, *Transversal homoclinic points and hyperbolic sets for nonautonomous maps*, Z.Angew.Math.Phys. **39** (1988), 518–549.

35. J.Waldvogel, *The period in the Lotka-Volterra-system predator prey model*, SIAM J.Num.Anal. **20** (1983), 186–192.

36. J.Waldvogel, *The period in the Lotka-Volterra system is monotinic*, J.of Math.Anal.and Appl. **114** (1986), 178–184.

INDEX

Vol. 1290: G. Wüstholz (Ed.), Diophantine Approximation and Transcendence Theory. Seminar, 1985. V, 243 pages. 1987.

Vol. 1291: C. Mœglin, M.-F. Vignéras, J.-L. Waldspurger, Correspondances de Howe sur un Corps p-adique. VII, 163 pages. 1987

Vol. 1292: J.T. Baldwin (Ed.), Classification Theory. Proceedings, 1985. VI, 500 pages. 1987.

Vol. 1293: W. Ebeling, The Monodromy Groups of Isolated Singularities of Complete Intersections. XIV, 153 pages. 1987.

Vol. 1294: M. Queffélec, Substitution Dynamical Systems – Spectral Analysis. XIII, 240 pages. 1987.

Vol. 1295: P. Lelong, P. Dolbeault, H. Skoda (Réd.), Séminaire d'Analyse P. Lelong – P. Dolbeault – H. Skoda. Seminar, 1985/1986. VII, 283 pages. 1987.

Vol. 1296: M.-P. Malliavin (Ed.), Séminaire d'Algèbre Paul Dubreil et Marie-Paule Malliavin. Proceedings, 1986. IV, 324 pages. 1987.

Vol. 1297: Zhu Y.-l., Guo B.-y. (Eds.), Numerical Methods for Partial Differential Equations. Proceedings. XI, 244 pages. 1987.

Vol. 1298: J. Aguadé, R. Kane (Eds.), Algebraic Topology, Barcelona 1986. Proceedings. X, 255 pages. 1987.

Vol. 1299: S. Watanabe, Yu. V. Prokhorov (Eds.), Probability Theory and Mathematical Statistics. Proceedings, 1986. VIII, 589 pages. 1988.

Vol. 1300: G.B. Seligman, Constructions of Lie Algebras and their Modules. VI, 190 pages. 1988.

Vol. 1301: N. Schappacher, Periods of Hecke Characters. XV, 160 pages. 1988.

Vol. 1302: M. Cwikel, J. Peetre, Y. Sagher, H. Wallin (Eds.), Function Spaces and Applications. Proceedings, 1986. VI, 445 pages. 1988.

Vol. 1303: L. Accardi, W. von Waldenfels (Eds.), Quantum Probability and Applications III. Proceedings, 1987. VI, 373 pages. 1988.

Vol. 1304: F.Q. Gouvêa, Arithmetic of p-adic Modular Forms. VIII, 121 pages. 1988.

Vol. 1305: D.S. Lubinsky, E.B. Saff, Strong Asymptotics for Extremal Polynomials Associated with Weights on ℝ. VII, 153 pages. 1988.

Vol. 1306: S.S. Chern (Ed.), Partial Differential Equations. Proceedings, 1986. VI, 294 pages. 1988.

Vol. 1307: T. Murai, A Real Variable Method for the Cauchy Transform, and Analytic Capacity. VIII, 133 pages. 1988.

Vol. 1308: P. Imkeller, Two-Parameter Martingales and Their Quadratic Variation. IV, 177 pages. 1988.

Vol. 1309: B. Fiedler, Global Bifurcation of Periodic Solutions with Symmetry. VIII, 144 pages. 1988.

Vol. 1310: O.A. Laudal, G. Pfister, Local Moduli and Singularities. V, 117 pages. 1988.

Vol. 1311: A. Holme, R. Speiser (Eds.), Algebraic Geometry, Sundance 1986. Proceedings. VI, 320 pages. 1988.

Vol. 1312: N.A. Shirokov, Analytic Functions Smooth up to the Boundary. III, 213 pages. 1988.

Vol. 1313: F. Colonius, Optimal Periodic Control. VI, 177 pages. 1988.

Vol. 1314: A. Futaki, Kähler-Einstein Metrics and Integral Invariants. IV, 140 pages. 1988.

Vol. 1315: R.A. McCoy, I. Ntantu, Topological Properties of Spaces of Continuous Functions. IV, 124 pages. 1988.

Vol. 1316: H. Korezlioglu, A.S. Ustunel (Eds.), Stochastic Analysis and Related Topics. Proceedings, 1986. V, 371 pages. 1988.

Vol. 1317: J. Lindenstrauss, V.D. Milman (Eds.), Geometric Aspects of Functional Analysis. Seminar, 1986–87. VII, 289 pages. 1988.

Vol. 1318: Y. Felix (Ed.), Algebraic Topology – Rational Homotopy. Proceedings, 1986. VIII, 245 pages. 1988

Vol. 1319: M. Vuorinen, Conformal Geometry and Quasiregular Mappings. XIX, 209 pages. 1988.

Vol. 1320: H. Jürgensen, G. Lallement, H.J. Weinert (Eds.), Semigroups, Theory and Applications. Proceedings, 1986. X, 416 pages. 1988.

Vol. 1321: J. Azéma, P.A. Meyer, M. Yor (Eds.), Séminaire de Probabilités XXII. Proceedings. IV, 600 pages. 1988.

Vol. 1322: M. Métivier, S. Watanabe (Eds.), Stochastic Analysis. Proceedings, 1987. VII, 197 pages. 1988.

Vol. 1323: D.R. Anderson, H.J. Munkholm, Boundedly Controlled Topology. XII, 309 pages. 1988.

Vol. 1324: F. Cardoso, D.G. de Figueiredo, R. Iório, O. Lopes (Eds.), Partial Differential Equations. Proceedings, 1986. VIII, 433 pages. 1988.

Vol. 1325: A. Truman, I.M. Davies (Eds.), Stochastic Mechanics and Stochastic Processes. Proceedings, 1986. V, 220 pages. 1988.

Vol. 1326: P.S. Landweber (Ed.), Elliptic Curves and Modular Forms in Algebraic Topology. Proceedings, 1986. V, 224 pages. 1988.

Vol. 1327: W. Bruns, U. Vetter, Determinantal Rings. VII,236 pages. 1988.

Vol. 1328: J.L. Bueso, P. Jara, B. Torrecillas (Eds.), Ring Theory. Proceedings, 1986. IX, 331 pages. 1988.

Vol. 1329: M. Alfaro, J.S. Dehesa, F.J. Marcellan, J.L. Rubio de Francia, J. Vinuesa (Eds.): Orthogonal Polynomials and their Applications. Proceedings, 1986. XV, 334 pages. 1988.

Vol. 1330: A. Ambrosetti, F. Gori, R. Lucchetti (Eds.), Mathematical Economics. Montecatini Terme 1986. Seminar. VII, 137 pages. 1988.

Vol. 1331: R. Bamón, R. Labarca, J. Palis Jr. (Eds.), Dynamical Systems, Valparaiso 1986. Proceedings. VI, 250 pages. 1988.

Vol. 1332: E. Odell, H. Rosenthal (Eds.), Functional Analysis. Proceedings, 1986–87. V, 202 pages. 1988.

Vol. 1333: A.S. Kechris, D.A. Martin, J.R. Steel (Eds.), Cabal Seminar 81–85. Proceedings, 1981–85. V, 224 pages. 1988.

Vol. 1334: Yu.G. Borisovich, Yu. E. Gliklikh (Eds.), Global Analysis – Studies and Applications III. V, 331 pages. 1988.

Vol. 1335: F. Guillén, V. Navarro Aznar, P. Pascual-Gainza, F. Puerta, Hyperrésolutions cubiques et descente cohomologique. XII, 192 pages. 1988.

Vol. 1336: B. Helffer, Semi-Classical Analysis for the Schrödinger Operator and Applications. V, 107 pages. 1988.

Vol. 1337: E. Sernesi (Ed.), Theory of Moduli. Seminar, 1985. VIII, 232 pages. 1988.

Vol. 1338: A.B. Mingarelli, S.G. Halvorsen, Non-Oscillation Domains of Differential Equations with Two Parameters. XI, 109 pages. 1988.

Vol. 1339: T. Sunada (Ed.), Geometry and Analysis of Manifolds. Procedings, 1987. IX, 277 pages. 1988.

Vol. 1340: S. Hildebrandt, D.S. Kinderlehrer, M. Miranda (Eds.), Calculus of Variations and Partial Differential Equations. Proceedings, 1986. IX, 301 pages. 1988.

Vol. 1341: M. Dauge, Elliptic Boundary Value Problems on Corner Domains. VIII, 259 pages. 1988.

Vol. 1342: J.C. Alexander (Ed.), Dynamical Systems. Proceedings, 1986–87. VIII, 726 pages. 1988.

Vol. 1343: H. Ulrich, Fixed Point Theory of Parametrized Equivariant Maps. VII, 147 pages. 1988.

Vol. 1344: J. Král, J. Lukeš, J. Netuka, J. Veselý (Eds.), Potential Theory – Surveys and Problems. Proceedings, 1987. VIII, 271 pages. 1988.

Vol. 1345: X. Gomez-Mont, J. Seade, A. Verjovski (Eds.), Holomorphic Dynamics. Proceedings, 1986. VII, 321 pages. 1988.

Vol. 1346: O. Ya. Viro (Ed.), Topology and Geometry – Rohlin Seminar. XI, 581 pages. 1988.

Vol. 1347: C. Preston, Iterates of Piecewise Monotone Mappings on an Interval. V, 166 pages. 1988.

Vol. 1348: F. Borceux (Ed.), Categorical Algebra and its Applications. Proceedings, 1987. VIII, 375 pages. 1988.

Vol. 1349: E. Novak, Deterministic and Stochastic Error Bounds in Numerical Analysis. V, 113 pages. 1988.

Vol. 1350: U. Koschorke (Ed.), Differential Topology. Proceedings, 1987. VI, 269 pages. 1988.

Vol. 1351: I. Laine, S. Rickman, T. Sorvali, (Eds.), Complex Analysis, Joensuu 1987. Proceedings. XV, 378 pages. 1988.

Vol. 1352: L.L. Avramov, K.B. Tchakerian (Eds.), Algebra – Some Current Trends. Proceedings, 1986. IX, 240 Seiten. 1988.

Vol. 1353: R.S. Palais, Ch.-l. Terng, Critical Point Theory and Submanifold Geometry. X, 272 pages. 1988.

Vol. 1354: A. Gómez, F. Guerra, M.A. Jiménez, G. López (Eds.), Approximation and Optimization. Proceedings, 1987. VI, 280 pages. 1988.

Vol. 1355: J. Bokowski, B. Sturmfels, Computational Synthetic Geometry. V, 168 pages. 1989.

Vol. 1356: H. Volkmer, Multiparameter Eigenvalue Problems and Expansion Theorems. VI, 157 pages. 1988.

Vol. 1357: S. Hildebrandt, R. Leis (Eds.), Partial Differential Equations and Calculus of Variations. VI, 423 pages. 1988.

Vol. 1358: D. Mumford, The Red Book of Varieties and Schemes. V, 309 pages. 1988.

Vol. 1359: P. Eymard, J.-P. Pier (Eds.), Harmonic Analysis. Proceedings, 1987. VIII, 287 pages. 1988.

Vol. 1360: G. Anderson, C. Greengard (Eds.), Vortex Methods. Proceedings, 1987. V, 141 pages. 1988.

Vol. 1361: T. tom Dieck (Ed.), Algebraic Topology and Transformation Groups. Proceedings, 1987. VI, 298 pages. 1988.

Vol. 1362: P. Diaconis, D. Elworthy, H. Föllmer, E. Nelson, G.C. Papanicolaou, S.R.S. Varadhan. École d'Été de Probabilités de Saint-Flour XV–XVII, 1985–87. Editor: P.L. Hennequin. V, 459 pages. 1988.

Vol. 1363: P.G. Casazza, T.J. Shura. Tsirelson's Space. VIII, 204 pages. 1988.

Vol. 1364: R.R. Phelps, Convex Functions, Monotone Operators and Differentiability. IX, 115 pages. 1989.

Vol. 1365: M. Giaquinta (Ed.), Topics in Calculus of Variations. Seminar, 1987. X, 196 pages. 1989.

Vol. 1366: N. Levitt, Grassmannians and Gauss Maps in PL-Topology. V, 203 pages. 1989.

Vol. 1367: M. Knebusch, Weakly Semialgebraic Spaces. XX, 376 pages. 1989.

Vol. 1368: R. Hübl, Traces of Differential Forms and Hochschild Homology. III, 111 pages. 1989.

Vol. 1369: B. Jiang, Ch.-K. Peng, Z. Hou (Eds.), Differential Geometry and Topology. Proceedings, 1986–87. VI, 366 pages. 1989.

Vol. 1370: G. Carlsson, R.L. Cohen, H.R. Miller, D.C. Ravenel (Eds.), Algebraic Topology. Proceedings, 1986. IX, 456 pages. 1989.

Vol. 1371: S. Glaz, Commutative Coherent Rings. XI, 347 pages. 1989.

Vol. 1372: J. Azéma, P.A. Meyer, M. Yor (Eds.), Séminaire de Probabilités XXIII. Proceedings. IV, 583 pages. 1989.

Vol. 1373: G. Benkart, J.M. Osborn (Eds.), Lie Algebras, Madison 1987. Proceedings. V, 145 pages. 1989.

Vol. 1374: R.C. Kirby, The Topology of 4-Manifolds. VI, 108 pages. 1989.

Vol. 1375: K. Kawakubo (Ed.), Transformation Groups. Proceedings, 1987. VIII, 394 pages. 1989.

Vol. 1376: J. Lindenstrauss, V.D. Milman (Eds.), Geometric Aspects of Functional Analysis. Seminar (GAFA) 1987–88. VII, 288 pages. 1989.

Vol. 1377: J.F. Pierce, Singularity Theory, Rod Theory, and Symmetry-Breaking Loads. IV, 177 pages. 1989.

Vol. 1378: R.S. Rumely, Capacity Theory on Algebraic Curves. III, 437 pages. 1989.

Vol. 1379: H. Heyer (Ed.), Probability Measures on Groups IX. Proceedings, 1988. VIII, 437 pages. 1989

Vol. 1380: H.P. Schlickewei, E. Wirsing (Eds.), Number Theory, U 1987. Proceedings. V, 266 pages. 1989.

Vol. 1381: J.-O. Strömberg, A. Torchinsky. Weighted Hardy Spac V, 193 pages. 1989.

Vol. 1382: H. Reiter, Metaplectic Groups and Segal Algebras. XI, pages. 1989.

Vol. 1383: D.V. Chudnovsky, G.V. Chudnovsky, H. Cohn, M.B. Nathan (Eds.), Number Theory, New York 1985–88. Seminar. V, 256 pages. 198

Vol. 1384: J. Garcia-Cuerva (Ed.), Harmonic Analysis and Pa Differential Equations. Proceedings, 1987. VII, 213 pages. 1989.

Vol. 1385: A.M. Anile, Y. Choquet-Bruhat (Eds.), Relativistic Fluid D mics. Seminar, 1987. V, 308 pages. 1989.

Vol. 1386: A. Bellen, C.W. Gear, E. Russo (Eds.), Numerical Method Ordinary Differential Equations. Proceedings, 1987. VII, 136 pages. 19

Vol. 1387: M. Petković, Iterative Methods for Simultaneous Inclusic Polynomial Zeros. X, 263 pages. 1989.

Vol. 1388: J. Shinoda, T.A. Slaman, T. Tugué (Eds.), Mathematical L and Applications. Proceedings, 1987. V, 223 pages. 1989.

Vol. 1000: Second Edition. H. Hopf, Differential Geometry in the Larg 184 pages. 1989.

Vol. 1389: E. Ballico, C. Ciliberto (Eds.), Algebraic Curves and Proj Geometry. Proceedings, 1988. V, 288 pages. 1989.

Vol. 1390: G. Da Prato, L. Tubaro (Eds.), Stochastic Partial Differ Equations and Applications II. Proceedings, 1988. VI, 258 pages. 19

Vol. 1391: S. Cambanis, A. Weron (Eds.), Probability Theory on V Spaces IV. Proceedings, 1987. VIII, 424 pages. 1989.

Vol. 1392: R. Silhol, Real Algebraic Surfaces. X, 215 pages. 1989.

Vol. 1393: N. Bouleau, D. Feyel, F. Hirsch, G. Mokobodzki Séminaire de Théorie du Potentiel Paris, No. 9. Proceedings. V pages. 1989.

Vol. 1394: T.L. Gill, W.W. Zachary (Eds.), Nonlinear Semigroups, Differential Equations and Attractors. Proceedings, 1987. IX, 233 1989.

Vol. 1395: K. Alladi (Ed.), Number Theory, Madras 1987. Proceedin 234 pages. 1989.

Vol. 1396: L. Accardi, W. von Waldenfels (Eds.), Quantum Probabi Applications IV. Proceedings, 1987. VI, 355 pages. 1989.

Vol. 1397: P.R. Turner (Ed.), Numerical Analysis and Parallel Proc Seminar, 1987. VI, 264 pages. 1989.

Vol. 1398: A.C. Kim, B.H. Neumann (Eds.), Groups – Korea 198 ceedings. V, 189 pages. 1989.

Vol. 1399: W.-P. Barth, H. Lange (Eds.), Arithmetic of Complex Ma Proceedings, 1988. V, 171 pages. 1989.

Vol. 1400: U. Jannsen. Mixed Motives and Algebraic K-Theory. X pages. 1990.

Vol. 1401: J. Steprāns, S. Watson (Eds.), Set Theory and its Appli Proceedings, 1987. V, 227 pages. 1989.

Vol. 1402: C. Carasso, P. Charrier, B. Hanouzet, J.-L. Joly Nonlinear Hyperbolic Problems. Proceedings, 1988. V, 249 page

Vol. 1403: B. Simeone (Ed.), Combinatorial Optimization. Semin V, 314 pages. 1989.

Vol. 1404: M.-P. Malliavin (Ed.), Séminaire d'Algèbre Paul Dubreil Paul Malliavin. Proceedings, 1987 – 1988. IV, 410 pages. 1989.

Vol. 1405: S. Dolecki (Ed.), Optimization. Proceedings, 1988 pages. 1989.

Vol. 1406: L. Jacobsen (Ed.), Analytic Theory of Continued Frac Proceedings, 1988. VI, 142 pages. 1989.

Vol. 1407: W. Pohlers, Proof Theory. VI, 213 pages. 1989.

Vol. 1408: W. Lück, Transformation Groups and Algebraic K-Th 443 pages. 1989.

Vol. 1409: E. Hairer, Ch. Lubich, M. Roche. The Numerical S Differential-Algebraic Systems by Runge-Kutta Methods. VII, 13 1989.

Vol. 1410: F.J. Carreras, O. Gil-Medrano, A.M. Naveira (Eds.), D Geometry. Proceedings, 1988. V, 308 pages. 1989.